IEE MONOGRAPH SERIES 19

Microwave attenuation measurement

D1356549

Microwave attenuation measurement

F.L. WARNER, C.Eng., F.I.E.E.
Senior Principal Scientific Officer
National R.F. and Microwave Standards Division
Royal Signals and Radar Establishment
Great Malvern
England

PETER PEREGRINUS LTD.
on behalf of the
Institution of Electrical Engineers

Published by Peter Peregrinus Ltd.,
Southgate House, Stevenage, Herts. SG1 1HQ, England

First published 1977
© 1977: Institution of Electrical Engineers

ISBN: 0 901223 79 4

Typeset at the Alden Press Oxford London and Northampton
Printed in England by A. Wheaton & Co. Ltd., Exeter

Preface

Over a hundred papers on microwave attenuation measurements have been published in the last 30 years. A few books devote one chapter to this very important subject, but most of them are out of date. In recent years, newcomers to microwave work have been almost bewildered by the large number of different techniques that can now be used to measure microwave attenuation. Thus, in the present book, the author has gathered together all the important information that has been published on this subject. Critical comparisons of the many different techniques are included, and several topics receive more detailed treatment here than in the original papers. Furthermore, this book contains some entirely new material that has not been published elsewhere.

This book has grown mainly from lecture notes that the author prepared for the IEE Vacation Schools on 'RF Electrical Measurements', which were held at the University of Surrey in 1973 and the University of Lancaster in 1976.

In spite of the wide and detailed coverage of this book, workers who have studied electronics to HNC level should have no difficulty in assimilating most of its contents. A list of the principal symbols used will be found after the Preface. The waveguide symbols are in accordance with British Standard 3939, Section 23.

As a result of two mergers, the name of the Establishment where the author works was changed, on 26th March 1976, from the Royal Radar Establishment (RRE) to the Royal Signals and Radar Establishment (RSRE). Where work by the author and his colleagues is mentioned, it is referred to as work at RRE if it was done before these mergers and as work at RSRE if carried out later.

I am extremely grateful to my former colleague, Ian Harris, (now retired) for making available hitherto unpublished work of his, which comprises almost all of Section 3.3, Chapter 14 and Appendices 3, 7 and 8. This last Appendix results from an interest taken by him in certain aspects of superconductivity and provides a conveniently compact background to the application of Josephson junctions to attenuation measurement.

I am also deeply indebted to Charles Ditchfield, who has a unique knowledge of microwave piston attenuators. He very kindly wrote Section 4.2 on this topic, specially for this book.

Charles Ditchfield, Ian Harris, Peter Herman and an unknown referee read through the entire manuscript and I am very grateful to them for suggesting many valuable improvements.

I greatly appreciate the help given to me by Mrs S.M. Randall, who typed more than half of the manuscript with remarkable accuracy. I am also deeply indebted to Mrs J. Bird, Mrs H.H.W. Bland, Miss R. Hawkey, Mrs S.A. Parkinson and Mrs J. Roberts, who typed the rest of the manuscript very competently.

Reproduction of Figs. 2.1, 2.8, 2.9, 3.6, 3.14, 4.1, 4.3, 4.4, 4.5, 4.7, 4.8, 4.10, 4.11, 4.12, 5.1, 5.2, 6.7, 7.3, 8.1, 8.3, 8.4, 8.5, 8.7, 8.8, 8.9, 8.10, 8.12, 8.13, 8.14, 9.3, 9.5, 9.6, 11.2, 11.3, 11.4, 11.5, 11.6, 11.7, 11.8, 11.9, 12.1, 12.4, 12.5, 12.6, 12.12, 13.7, 13.8, 13.9, A1.1, A2.1, A2.3, and A2.4 is by permission of the Controller HMSO, London, 1976, and Crown Copyright is reserved. Figs. 4.10, 4.11, 4.12, 8.12, 8.13, 8.14, 9.3, 9.5 and 9.6 have previously appeared in IEEE publications. Permission to use them again is gratefully acknowledged.

Finally, I thank my wife for being so patient throughout the many evenings I spent writing this book.

Malvern
October 1976

F.L. Warner

Contents

Principal symbols

The popular symbols are used with numerous different subscripts, which are clearly defined in the text. To prevent this list becoming unwieldy, these popular symbols are only given once with an asterisk in the subscript position. SI units are used throughout.

A_* = attenuation, dB
A_c = cross-sectional area of toroidal core (in Chap. 3)
B_* = magnetic flux density
B = susceptance of waveguide step (in Chap. 11)
C_* = capacitance
C = coupling parameter of directional coupler shown in Fig. 13.5
D_* = setting of a Kelvin-Varley or inductive voltage divider
D = denominator obtained when using Mason's non-touching loop rule
E_* = electric field strength
F = noise factor (expressed as a power ratio)
G_* = conductance
G_1, G_2 = gains of receiving systems discussed in Section 12.9
H_* = magnetic field strength
I_* = electric current
J = bolometer constant
$J_n(x)$ = Bessel function of the first kind and nth order
K_* = constant
L_* = inductance
L_I = insertion loss, dB
L_s = substitution loss, dB
L_T = transducer loss, dB
L_v = voltage loss, dB

a = radius (in Chap. 3)

a = complex quantity in the bilinear transformation equation (in Chap. 9)

a = larger internal dimension of rectangular waveguide (in Chap. 13)

a_1, a_2 etc. = complex entering wave amplitudes

a_i, a_0 = inner and outer radii, respectively, of a toroidal core

b = width of toroidal core (in Chap. 3)

b = complex quantity in the bilinear transformation equation (in Chap. 9)

b = normalised susceptance of a waveguide step (in Chapter 11)

b_1, b_2 etc. = complex outgoing wave amplitudes

c = mean circumference of a toroidal core (in Chap. 3)

c = speed of electromagnetic waves in free space $= 2 \cdot 997925 \times 10^8 \, \text{m/s}$

c = complex quantity in the bilinear transformation equation (in Chap. 9)

d = diameter of wire (in Chap. 3)

d'' = diameter of 10 strand cable (in Chap. 3)

d = internal diameter of piston attenuator bore (in Chapter 4)

d = complex quantity in the bilinear transformation equation (in Chap. 9)

d = directivity of directional coupler, dB (in Chap. 10)

d = distance shown in Fig. 12.8b

e = base of natural logarithms $= 2 \cdot 718281828$

e = charge on electron $= -1 \cdot 60219 \times 10^{-19} \, \text{C}$

e_* = signal amplitude

e_0, e_1, e_2 = constants of an automatic network analyser that are determined by the calibration procedure

f_* = frequency

h = Planck's constant $= 6 \cdot 6262 \times 10^{-34} \, \text{J s}$

h_n = smaller internal dimension of normal waveguide

h_r = smaller internal dimension of reduced height waveguide

i_* = electric current

j = $\sqrt{-1}$

k = Boltzmann's constant $= 1 \cdot 38063 \times 10^{-23} \, \text{J/deg K}$

k = ρ_{nm}/a (in Chap. 3)

k = complex quantity in bilinear transformation equation (in Chap. 9)

M_1 = mismatch loss factor between generator and 2-port (in Chap. 2)

M_2 = mismatch loss factor between 2-port and load (in Chap. 2)

M_I = mismatch error when measuring attenuation

M_s = mismatch error when measuring incremental attenuation

M_* = mutual inductance (in Chap. 3)

M = quantity defined in eqn. 4.13

M_* = measured voltage ratio (in Chap. 11)

N = total number of turns on a toroidal core (in Chap. 3)

N_b = base of number system

N = quantity defined in eqn. 4.14

P_* = power

Q = quality factor of resonant cavity

R_* = resistance

R_{pa} = attenuation rate of a microwave piston attenuator (in Chap. 4)

R = radius of transformed circle (in Chap. 9)

S_* = voltage standing wave ratio

T = transformer turns ratio (in Chap. 3)

T = constant in eqn. 8.1

T_0 = standard reference temperature (290 K)

T = transmission parameter of the directional coupler shown in Fig. 13.5

U = constant in eqn. 8.1

U = uncertainty (in Chap. 14)

U_r = random component of uncertainty

U_s = systematic component of uncertainty

V_* = voltage

W_0, W_z, W_a = d.c. bias powers dissipated in a bolometer when the microwave input powers are 0, P_z and P_a, respectively

W = distance from the origin to the centre of the transformed circle (in Chap. 9)

W = width of the intrinsic region of an impatt diode (in Chap. 10)

X_* = reactance

Y_* = admittance

Z_* = impedance

Z_0 = characteristic impedance of transmission line or waveguide

$$k = V_{B1}/V_A \text{ (in Chap. 12)}$$
$$k = \{\text{antilog } (A/20)\}^{-1} \text{ (in Chap. 13)}$$
$$k_0 = \text{coupling coefficient}$$
$$l_* = \text{length}$$
$$l_{11}, l_{12}, l_{21}, l_{22} = \text{scattering parameters of a 2-port network}$$
$$\bar{l} = \text{mean leakage inductance of one section of an IVD}$$
$$m = \text{modulation depth}$$
$$m_{11}, m_{12}, m_{13} \text{ etc.} = \text{scattering parameters}$$
$n =$ number of sections between the tap and earth in a 10 section autotransformer (in Chap. 3)

$n =$ number of turns required on the worm shaft to move the central vane of a rotary attenuator through $90°$ (in Chap. 4)

$n =$ bolometer constant (in Chap. 8)

$n =$ number of times a measurement is repeated (in Chap. 14)

$$n_{11}, n_{12}, n_{13} \text{ etc.} = \text{scattering parameters}$$
$$o_{11}, o_{12}, o_{21}, o_{22} = \text{scattering parameters shown in Figs. 11.6 and 11.7}$$
$$p = \text{r.f. input power to bolometer}$$
$$p_{11}, p_{12}, p_{21}, p_{22} = \text{scattering parameters shown in Fig. 11.6}$$
$q =$ ratio of H_{12} mode amplitude to H_{11} mode amplitude at $z = 0$ in a piston attenuator

$$q_1, q_2 = \text{scattering parameters shown in Figs. 11.6 and 11.7}$$
$$r = \text{radial distance in cylindrical coordinate system}$$
$$r = \frac{E}{V}(1 + m \sin \omega_m t) \text{ in Chap. 8}$$
$$r_{s1}, r_{s2}, r_{s3} = \text{lead and contact resistances in a Kelvin-Varley divider}$$
$$r_t = \text{total series resistance of a toroidal autotransformer}$$
$$\bar{r} = r_t/10$$
$$r_1, r_2 = \text{scattering parameters shown in Fig. 11.6}$$
$s =$ reflection coefficient of the short-to-match switch shown in Fig. 13.5

$$s_* = \text{standard deviation of a set of measurements}$$
$$s_{11}, s_{12}, s_{21}, s_{22} = \text{scattering parameters of a 2-port network}$$
s_{11b}, s_{12b} etc. $=$ scattering parameters of a variable 2-port at the beginning of a substitution loss measurement

s_{11e}, s_{12e} etc. $=$ scattering parameters of a variable 2-port at the end of a substitution loss measurement

$$t = \text{lamination thickness (in Chap. 3)}$$
$t =$ thickness of epitaxial layer in a Gunn diode (in Chap. 10)

$$t = \frac{o_{21}q_2}{1 - o_{22}q_1} \text{ (in Chap. 11)}$$

t = 'Student-t' factor (in Chap. 14)

$$u = \frac{p_{21}r_2}{1 - p_{22}r_1}$$

v = domain velocity in a Gunn diode

v_s = velocity at which the holes travel through the intrinsic region of an impatt diode

$v_{11}, v_{12}, v_{21}, v_{22}$ = scattering parameters shown in Fig. 11.7

w = thickness of the casing around a toroidal core

w_1, w_2 = scattering parameters shown in Fig. 11.7

x = real part of Γ_G (in Chap. 2)

x' = real part of Γ_2 (in Chap. 2)

$$x = \frac{v_{21}w_2}{1 - v_{22}w_1} \text{(in Chap. 11)}$$

x = fraction of saw-tooth voltage that is applied to the X amplifier (in Chap. 12)

x = distance from side of waveguide to vane (in Chap. 13)

$\bar{x}$ = mean value obtained in a set of measurements (in Chap. 14)

x_i = ith result obtained in a set of measurements (in Chap. 14)

y = imaginary part of Γ_G (in Chap. 2)

y' = imaginary part of Γ_2 (in Chap. 2)

y = $|\Gamma|e^{-2\alpha_m l}$ (in Chap. 12)

z = axial distance in cylindrical coordinate system

z_1, z_2, z_3 = series impedances shown in Fig. 3.16b

Γ_* = reflection coefficient

$\Delta_1, \Delta_2, \Delta_3$ etc. = deviations from nominal resistance values in binary Kelvin-Varley divider

ΔA_* = error or uncertainty in attenuation measurement

Λ = quantity defined in eqn 11.31

Σ = sum of

Φ = magnetic flux linking a SQUID

Φ_a = component of Φ due to an external source

Φ_0 = flux quantum = $h/2e = 2 \cdot 0678538 \times 10^{-15}$ W

α_* = attenuation constant in Np/m

α = h_r/h_n (in Chap. 11)

α_{nm} = attenuation constant for the E_{nm} or H_{nm} mode

β = phase constant = $2\pi/\lambda$ or $2\pi/\lambda_g$

γ = ratio of local oscillator voltage to maximum signal voltage applied to mixer, dB

δ = skin depth

ϵ = half of angle between the end vanes in a rotary attenuator

ϵ = permittivity = $\epsilon_r\epsilon_0$

ϵ_r = permittivity relative to a vacuum

ϵ_0 = $8\cdot85419 \times 10^{-12}$ F/m

$\epsilon_{oz}, \epsilon_{oa}, \epsilon_{az}$ = fractional systematic errors that occur when power differences are measured

ζ = total error in an IVD

ζ_3 = 3 terminal linearity of Kelvin-Varley divider

ζ_4 = 4 terminal linearity of Kelvin-Varley divider

ζ_{abs} = Absolute linearity of Kelvin-Varley divider

ζ_* = one component of ζ

η = network efficiency (in Chap. 2)

η = ratio of i.f. output voltage to signal input voltage in linear mixer (in Chap. 8)

θ_* = phase angle or phase difference

θ = angle between central vane and end vanes in rotary attenuator (in Chap. 4)

λ = free space wavelength

λ_c = cut-off wavelength

λ_g = guide wavelength

μ = permeability = $\mu_r\mu_0$

μ_r = permeability relative to a vacuum

μ_0 = $4\pi \times 10^{-7}$ H/m

μ_t = relative permeability of a toroidal core

ξ = hysteresis loss/m^3/Hz in a toroidal core (in Chap. 3)

ξ = differential phase change between the components parallel and perpendicular to the central vane in a rotary attenuator (in Chap. 4)

ξ = $Em_{31}t$ (in Chap. 11)

ξ = phase change that occurs on reflection from the metal partition in the attenuation transfer standard shown in Fig. 13.5

π = $3\cdot1415926536$

ρ_* = resistivity

ρ_{nm} = mth zero of nth-order Bessel function $J_n(x)$ for E modes or mth zero of the derivative of the nth-order Bessel function $J'_n(x)$ for H modes

σ_* = conductivity

τ = available power transmission factor (in Chap. 2)

τ = output time constant (in Chap. 8)

ϕ = angle in cylindrical coordinate system

ϕ_* = phase angle or phase difference

ψ_* = phase angle or phase difference

ω_* = $2\pi f_*$

Chapter 1

Introduction

When a microwave signal is sent along a uniform lossy transmission line, which is matched* at the far end, the signal amplitude decreases exponentially with distance, and the signal is said to be attenuated. In a practical microwave system, the situation is more complex. Let us consider a radar transmitter which is connected to an aerial with a waveguide system containing a duplexer and a rotating joint. Small discontinuities will be present at various points in the waveguide system and the transmitter and aerial will scarcely ever be perfectly matched, so multiple reflections will occur. Both the forward and backward waves will be attenuated by losses in the waveguide walls and by any lossy elements that are present in the duplexer and rotating joint. Some of the reflected power will be dissipated in the transmitter. The forward and backward waves will interact with each other to produce standing wave patterns inside the waveguide. Internal reflections usually reduce the radiated power, but, if they produce a conjugate match (see Section 2.4) between the transmitter and the aerial, the radiated power may be increased beyond the value obtained when the duplexer, rotating joint and aerial are all perfectly matched to the characteristic impedance of the waveguide. With all of these complications, a precise definition of attenuation is clearly essential. Attenuation is nearly always expressed in decibels, which is abbreviated to dB, or in nepers.

The attenuation A, in dB, of a 2-port device is defined as follows:

$$A = 10 \log_{10} \left\{ \frac{\text{power delivered to a matched load by a matched generator which is connected directly to it}}{\text{power delivered to the same load by the same generator when the 2-port device is inserted between them}} \right\}$$

$$(1.1)$$

* Throughout this book, unless otherwise stated, the term 'matched' means matched to the characteristic impedance of the transmission system

Attenuation in nepers $= \frac{1}{2} \log_e$ (power ratio in eqn. 1.1)   (1.2)

As $\log_{10} x = \log_e x / \log_e 10$, it is seen that:

Attenuation in decibels $= 8 \cdot 6858896 \times$ attenuation in nepers

(1.3)

In addition to attenuation, many other terms are used in the subject of loss measurement and these are precisely defined in Chapter 2.

Applying eqn. 1.1 to the radar system considered earlier, it can be seen that a total attenuation of 2 dB in the waveguide, duplexer and rotating joint causes 37% of the transmitter power to be wasted if both the transmitter and aerial are perfectly matched. Thus, in a radar system, it is a matter of paramount importance to reduce all of the losses in the waveguide system to an absolute minimum. Attenuation measurements are therefore frequently essential when microwave components are being developed and tested for use in radar systems.

Attenuation is equally important in the case of reception. At a satellite receiving station, extremely weak signals have to be picked up, so a maser is sometimes used at the front end of the receiver. Such a device has a noise temperature of less than 5 K, but, if it is connected to the aerial with a length of uncooled waveguide that has a loss of only $0 \cdot 1$ dB in it, the noise temperature is raised by 7 K. Thus, very precise attenuation measurements should be carried out on any lengths of waveguide that are made for such a system.

To check the noise temperature of a satellite receiving system, a liquid helium or liquid nitrogen cooled noise standard is used, and the attenuation in all parts of the waveguide in such a device must be measured very accurately to determine the effective noise temperature at the output flange. A neglected loss of $0 \cdot 01$ dB in the output section of a liquid nitrogen noise standard causes a 1% error in the noise temperature.

Similar needs for precise attenuation measurements exist throughout the entire microwave field. Just one more example will be given. A lot of work is being done on long-distance millimetre-wave communication systems, which use the low-loss H_{01} mode in helical waveguides of circular cross-section./Accurate attenuation measurements have to be made on the very long lengths of waveguide which are made for these systems before they are buried deeply in the ground.

Many different methods are used to measure microwave attenuation[1-9] and they can be grouped as follows:

(*a*) power ratio methods
(*b*) comparison with an accurate microwave attenuator (r.f.

substitution)

(*c*) comparison with an intermediate frequency piston attenuator (i.f. substitution)

(*d*) comparison with an audio frequency attenuator (a.f. substitution)

(*e*) methods based on reflection coefficient measurements

(*f*) swept frequency methods

(*g*) methods using microwave network analysers

(*h*) techniques that rely on the addition and subtraction of signals from 3 different paths

(*i*) twin-channel null and off-null techniques

(*j*) shuttle pulse method

(*k*) methods based on Q measurements

(*l*) method using a Josephson junction in a superconducting loop

(*m*) miscellaneous methods

Methods (*a*) to (*g*) are described in Chapters 5 to 11, respectively, and the other methods are described in Chapter 12. In some of the methods listed above, it is seen that the attenuation standard operates at d.c., a.f. or i.f. and it is then usually a Kelvin Varley voltage divider, an inductive voltage divider or an i.f. piston attenuator. These three devices are described in Chapter 3. The r.f. substitution method requires an attenuation standard which operates at the same frequency as the device under test. Two devices that are suitable for this role are the waveguide rotary vane attenuator and the microwave piston attenuator. Both of these devices are fully discussed in Chapter 4.

Attenuation transfer standards and the assessment of uncertainty in attenuation measurements are the topics dealt with in Chapters 13 and 14. In the last chapter, the advantages and disadvantages of all the well established methods for measuring microwave attenuation are critically compared and possible future developments are mentioned.

The first and second appendices give self-contained treatments of scattering parameters and signal flow graphs. Evanescent electromagnetic waves inside perfectly conducting cylinders are analysed fully in Appendix 3. Four further appendices give mathematical derivations of equations that are used in the main text. The last appendix is devoted to the theory of weakly connected superconducting rings. A list of the principal symbols used in this book is found after the Preface.

References

1 MONTGOMERY, C.G.: 'Technique of microwave measurements' (McGraw-Hill, 1947), chap. 13

2 BARLOW, H.M., and CULLEN, A.L.: 'Microwave measurements' (Constable, 1950), chap. 8

3 WIND, M.: 'Handbook of electronic measurements – Vol. 2' (Polytechnic Press, 1956), chap. 18

4 GINZTON, E.L.: 'Microwave measurements' (McGraw-Hill, 1957), chap. 11

5 SUCHER, M., and FOX, J.: 'Handbook of microwave measurements – Vol. 1' (Polytechnic Press, 1963), chap. 7

6 EBERT, J.E., and SORGER, G.U.: 'Survey of precision microwave attenuation measuring techniques', *ISA Trans.,* 1964, **3**, pp. 280–290

7 BEATTY, R.W.: 'Microwave attenuation measurements and standards'. NBS Monograph 97, April 1967

8 RUSSELL, D., and LARSON, W.: 'RF attenuation', *Proc. IEEE,* 1967, **55**, pp. 942–959

9 GASKELL, C.S.: 'Microwave attenuation measurement', *Design Electronics,* 1969, **6**, pp. 30–33

Chapter 2

Basic definitions and equations related to attenuation

In the subject of loss measurement, many different terms are used.[1,2] These include: insertion loss, attenuation, reflective component of attenuation, dissipative component of attenuation, substitution loss, residual attenuation, incremental attenuation, transducer loss, voltage loss, network efficiency, available power transmission factor and mismatch loss factor. In this chapter, all of these terms are defined and mathematical expressions for them are derived from first principles, using scattering parameters and signal flow graphs. Any reader who is not already familiar with scattering parameters and signal flow graphs will find self-contained treatments of these two topics in Appendices 1 and 2. With any method of attenuation measurement, errors arise from leakage and imperfect matching at the insertion point. Theoretical expressions for these errors are derived in Sections 2.11 and 2.12.

Throughout this chapter, it is assumed that the input and output lines have the same characteristic impedance. Beatty[1,2] has given equations that remain valid when the characteristic impedances are unequal.

2.1 Insertion loss and attenuation

When a generator with a reflection coefficient Γ_G is connected directly to a load with a reflection coefficient Γ_L, let the power dissipated in the load be denoted by P_1. Suppose now that a 2-port network is connected between the generator and the load, as shown in Fig. 2.1, and let this reduce the power dissipated in the load to P_2. Then, by definition, the insertion loss of this 2-port network is given in decibels by

$$L_I = 10 \log_{10} \frac{P_1}{P_2} \tag{2.1}$$

Attenuation is defined as the insertion loss when $\Gamma_G = 0$ and $\Gamma_L = 0$. Insertion loss depends on the values of Γ_G and Γ_L as well as on the characteristics of the 2-port network, whereas attenuation is a property only of the 2-port network.

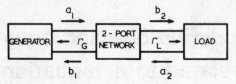

Fig. 2.1 2-port network between generator and load

Using scattering coefficients s_{11}, s_{12}, s_{21} and s_{22} to define the 2-port network, the complex wave amplitudes shown in Fig. 2.1 are related as follows[*]:

$$b_1 = s_{11}a_1 + s_{12}a_2 \tag{2.2}$$

$$b_2 = s_{21}a_1 + s_{22}a_2 \tag{2.3}$$

The signal flow graph for the configuration given in Fig. 2.1 is shown in Fig. 2.2. Using the non-touching loop rule[†], we immediately get

$$\frac{b_2}{e} = \frac{s_{21}}{1 - (\Gamma_G s_{11} + s_{22}\Gamma_L + \Gamma_G s_{21}\Gamma_L s_{12}) + \Gamma_G s_{11} s_{22}\Gamma_L} \tag{2.4}$$

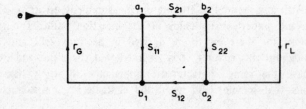

Fig. 2.2 Signal flow graph for the configuration shown in Fig. 2.1

Denoting the assumed real characteristic impedance[*] of the transmission system by Z_0, the power incident on the load is $|b_2|^2/Z_0$ and the power reflected from the load is $|a_2|^2/Z_0$. Thus, the power dissipated in the load P_2 is as follows:

$$P_2 = \frac{|b_2|^2}{Z_0} - \frac{|a_2|^2}{Z_0} = \frac{|b_2|^2}{Z_0}(1 - |\Gamma_L|^2) \tag{2.5}$$

[*] Precise definitions of a_1, a_2, b_1, b_2, e and Z_0 are given in Appendix 1
[†] When using the non-touching loop rule, the scattering parameters have been written down in the order in which they occur as one moves around the various paths and loops, starting wherever possible on the left-hand side

Inserting eqn. 2.4 into 2.5 and rearranging the denominator, we get

$$P_2 = \frac{|e|^2 |s_{21}|^2 (1 - |\Gamma_L|^2)}{Z_0 |(1 - \Gamma_G s_{11})(1 - \Gamma_L s_{22}) - \Gamma_G \Gamma_L s_{12} s_{21}|^2} \qquad (2.6)$$

The power P_1 dissipated in the load when the generator is connected directly to it can be found immediately from eqn. 2.6 by letting $s_{11} = s_{22} = 0$ and $s_{12} = s_{21} = 1$.

Thus $$P_1 = \frac{|e|^2 (1 - |\Gamma_L|^2)}{Z_0 |1 - \Gamma_G \Gamma_L|^2} \qquad (2.7)$$

Substituting eqns. 2.6 and 2.7 into eqn. 2.1, we get

$$L_I = 10 \log_{10} \frac{|(1 - \Gamma_G s_{11})(1 - \Gamma_L s_{22}) - \Gamma_G \Gamma_L s_{12} s_{21}|^2}{|s_{21}|^2 |1 - \Gamma_G \Gamma_L|^2} \qquad (2.8)$$

which can be simplified further to

$$L_I = 20 \log_{10} \frac{|(1 - \Gamma_G s_{11})(1 - \Gamma_L s_{22}) - \Gamma_G \Gamma_L s_{12} s_{21}|}{|s_{21}| \cdot |1 - \Gamma_G \Gamma_L|} \qquad (2.9)$$

An expression for the attenuation A of the 2-port can now be found straight away from eqn. 2.9 by setting both Γ_G and Γ_L equal to zero.

Hence $$A = 20 \log_{10} \frac{1}{|s_{21}|} \qquad (2.10)$$

2.2 Reflective and dissipative components of attenuation

Attenuation is sometimes separated into two components, one associated with reflection and the other with dissipation. Let

P_I = power incident on the 2-port

P_R = power reflected by the 2-port back into the matched generator

P_L = power dissipated in the matched load.

Then, the reflective component of the attenuation is given by

$$A_r = 10 \log_{10} \frac{P_I}{P_I - P_R} \qquad (2.11)$$

and the dissipative component of the attenuation is given by

$$A_d = 10 \log_{10} \frac{P_I - P_R}{P_L} \qquad (2.12)$$

In this case, where both the generator and load are matched,

$$\frac{P_R}{P_I} = |s_{11}|^2 \quad \text{and} \quad \frac{P_L}{P_I} = |s_{21}|^2 \qquad (2.13), (2.14)$$

Therefore:

$$A_r = 10 \log_{10} \frac{1}{1 - |s_{11}|^2} \quad \text{and} \quad A_d = 10 \log_{10} \frac{1 - |s_{11}|^2}{|s_{21}|^2}$$

$$(2.15), (2.16)$$

Clearly $$A = A_r + A_d \qquad (2.17)$$

This last equation assumes great importance in connection with the reflection coefficient methods of measuring attenuation which are described in Chapter 9.

2.3 Substitution loss

Let us now consider a slightly different case. At the beginning, let us suppose that a 2-port network with scattering parameters s_{11b}, s_{12b}, s_{21b} and s_{22b} is connected between the source and the load shown in Fig. 2.1. Then, the power P_3 delivered to the load is seen straight away from eqn. 2.6 to be

$$P_3 = \frac{|e|^2 |s_{21b}|^2 (1 - |\Gamma_L|^2)}{Z_0 |(1 - \Gamma_G s_{11b})(1 - \Gamma_L s_{22b}) - \Gamma_G \Gamma_L s_{12b} s_{21b}|^2} \qquad (2.18)$$

Let us now suppose that this 2-port is removed and another one with scattering parameters s_{11e}, s_{12e}, s_{21e} and s_{22e} is substituted in its place. The power P_4 that is now delivered to the load is given by

$$P_4 = \frac{|e|^2 |s_{21e}|^2 (1 - |\Gamma_L|^2)}{Z_0 |(1 - \Gamma_G s_{11e})(1 - \Gamma_L s_{22e}) - \Gamma_G \Gamma_L s_{12e} s_{21e}|^2} \qquad (2.19)$$

Then, by definition, the substitution loss is given in decibels by

$$L_s = 10 \log_{10} \frac{P_3}{P_4} \qquad (2.20)$$

From eqns. 2.18, 2.19 and 2.20 we get

$$L_s = 20 \log_{10} \frac{|s_{21b}| \cdot |(1 - \Gamma_G s_{11e})(1 - \Gamma_L s_{22e}) - \Gamma_G \Gamma_L s_{12e} s_{21e}|}{|s_{21e}| \cdot |(1 - \Gamma_G s_{11b})(1 - \Gamma_L s_{22b}) - \Gamma_G \Gamma_L s_{12b} s_{21b}|}$$

$$(2.21)$$

If the first 2-port is a perfect connector pair with $s_{11b} = s_{22b} = 0$ and $s_{12b} = s_{21b} = 1$, then it can be seen from eqns. 2.21 and 2.9 that the substitution loss becomes equal to the insertion loss. When $\Gamma_G = \Gamma_L = 0$, the substitution loss is given by

$$A_{inc} = 20 \log_{10} \frac{1}{|s_{21e}|} - 20 \log_{10} \frac{1}{|s_{21b}|} \qquad (2.22)$$

and is clearly equal to the difference between the attenuations through the final and initial 2-port networks.

Substitution loss is the parameter that is measured when a variable microwave attenuator is calibrated. When such a device is set to its 'zero' or 'datum' position, a certain amount of attenuation remains in it and this is called the *residual attenuation*. In a waveguide rotary attenuator, or a sideways displacement attenuator, a typical value for the residual attenuation is 0·15 dB, but, in a microwave piston attenuator, it may exceed 30 dB. When a variable attenuator is moved from its 'zero' or 'datum' position to any other setting, the change in attenuation that occurs is called the *incremental attenuation*. Moving a variable attenuator in this way is equivalent to starting with one 2-port network and then substituting another one in its place. Thus, if $|s_{21b}|$ corresponds to the 'zero' or 'datum' setting, eqn. 2.22 gives the incremental attenuation.

2.4 Transducer loss

It is well known, from elementary circuit theory, that a generator is able to deliver maximum power to a load, which is connected directly to it, when the load impedance is equal to the complex conjugate of the generator impedance. In an equivalent microwave configuration with a real characteristic impedance, it is easily shown that the above condition is satisfied when the load reflection coefficient is equal to the complex conjugate of the generator reflection coefficient. Thus, let $\Gamma_G = x + jy$ and let Γ_L be transformed by a completely lossless matching unit to $x - jy$. Then, on substituting these expressions into eqn. 2.7, the maximum power that can be delivered to the load is found to be

$$P_{cm} = \frac{|e|^2}{Z_0(1 - |\Gamma_G|^2)} \qquad (2.23)$$

P_{cm} is equal to the 'available power' from the generator. By definition, the transducer loss expressed in decibels is given by

$$L_T = 10 \log_{10} \frac{P_{cm}}{P_2} \tag{2.24}$$

From eqns. 2.6, 2.23 and 2.24 we find that

$$L_T = 10 \log_{10} \frac{|(1 - \Gamma_G s_{11})(1 - \Gamma_L s_{22}) - \Gamma_G \Gamma_L s_{12} s_{21}|^2}{|s_{21}|^2 (1 - |\Gamma_G|^2)(1 - |\Gamma_L|^2)} \tag{2.25}$$

Hence, transducer loss is equal to the substitution loss when the first 2-port network is a completely lossless transducer which provides conjugate matching of the load to the generator. Transducer loss is an interesting concept and has received a lot of attention in the literature. However, it is rarely used in practice as the very best matching devices are slightly lossy. Thus, after achieving a conjugate match and making a measurement of the apparent transducer loss, it is necessary to measure, separately, the loss in the matching unit and then add this result to the previous one.

2.5 Voltage loss

The voltage loss L_v of a 2-port network is defined as follows:

$$L_v = 20 \log_{10} \left\{ \frac{\text{voltage at input of 2-port}}{\text{voltage at output of 2-port}} \right\} \tag{2.26}$$

Thus, referring again to Fig. 2.1, we see that

$$L_v = 20 \log_{10} \frac{|a_1 + b_1|}{|a_2 + b_2|} \tag{2.27}$$

Using the non-touching loop rule, we immediately get

$$\frac{a_1}{e} = \frac{1 - s_{22}\Gamma_L}{D} \tag{2.28}$$

$$\frac{b_1}{e} = \frac{s_{11}(1 - s_{22}\Gamma_L) + s_{21}\Gamma_L s_{12}}{D} \tag{2.29}$$

$$\frac{a_2}{e} = \frac{s_{21}\Gamma_L}{D} \tag{2.30}$$

$$\frac{b_2}{e} = \frac{s_{21}}{D} \tag{2.31}$$

where

$$D = 1 - (\Gamma_G s_{11} + s_{22}\Gamma_L + \Gamma_G s_{21}\Gamma_L s_{12}) + \Gamma_G s_{11}s_{22}\Gamma_L \quad (2.32)$$

Substitution of eqns. 2.28 to 2.31 into eqn. 2.27 yields

$$L_v = 20\log_{10}\frac{|(1 + s_{11})(1 - s_{22}\Gamma_L) + s_{21}\Gamma_L s_{12}|}{|s_{21}(1 + \Gamma_L)|} \quad (2.33)$$

It should be noted that L_v is quite independent of Γ_G. It can be seen from eqn. 2.33 that the voltage loss becomes equal to the attenuation when $s_{11} = 0$ and $\Gamma_L = 0$, i.e. when perfect matches are seen looking both directly into the load and into the 2-port when it is followed by a matched load. Voltage loss is a quantity that is normally only used when making measurements in the d.c. to u.h.f. part of the spectrum.

In the past some workers have confused voltage loss with insertion loss. If one wishes to work in terms of voltages, the correct definition for insertion loss is

$$L_I = 20\log_{10}\left\{\frac{\begin{array}{c}\text{voltage across load when it is}\\ \text{connected directly to the generator}\end{array}}{\text{voltage at output of 2-port}}\right\} \quad (2.34)$$

The voltage across the load with a direct connection to the generator is equal to $|a_2 + b_2|$ when $s_{11} = s_{22} = 0$ and $s_{12} = s_{21} = 1$. Thus

$$L_I = 20\log_{10}\frac{\left|\dfrac{e\Gamma_L}{1 - \Gamma_G\Gamma_L} + \dfrac{e}{1 - \Gamma_G\Gamma_L}\right|}{\left|\dfrac{es_{21}\Gamma_L}{D} + \dfrac{es_{21}}{D}\right|} \quad (2.35)$$

which, after rearrangement, becomes identical with eqn. 2.9.

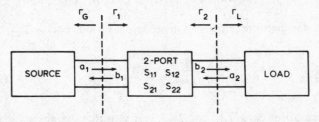

Fig. 2.3 Diagram defining symbols used in Sections 2.6, 2.7 and 2.8

2.6 Network efficiency

In papers on the measurement of microwave noise,[3-5] three further loss terms are used. These are: network efficiency, available power transmission factor and mismatch loss factor. For completeness, expressions for these three quantities will now be derived. Referring to Fig. 2.3, the network efficiency η of the 2-port circuit is defined as follows:

$$\eta = \frac{\text{power absorbed by load}}{\text{power absorbed at inlet of 2-port}} \qquad (2.36)$$

$$= \frac{\dfrac{|b_2|^2}{Z_0} - \dfrac{|a_2|^2}{Z_0}}{\dfrac{|a_1|^2}{Z_0} - \dfrac{|b_1|^2}{Z_0}} \qquad \text{(assuming that } Z_0 \text{ is real)} \qquad (2.37)$$

$$= \frac{|b_2|^2}{|a_1|^2} \cdot \frac{\{1 - |\Gamma_L|^2\}}{\{1 - |\Gamma_1|^2\}} \qquad (2.38)$$

where

$$\Gamma_1 = s_{11} + \frac{s_{21}\Gamma_L s_{12}}{1 - s_{22}\Gamma_L} \qquad (2.39)$$

From eqns. 2.28 and 2.31, we get

$$\frac{b_2}{a_1} = \frac{s_{21}}{1 - s_{22}\Gamma_L} \qquad (2.40)$$

Substitution of eqn. 2.40 into 2.38 yields:

$$\eta = \frac{|s_{21}|^2 \{1 - |\Gamma_L|^2\}}{|1 - s_{22}\Gamma_L|^2 \{1 - |\Gamma_1|^2\}} \qquad (2.41)$$

$$= \frac{|s_{21}|^2 \{1 - |\Gamma_L|^2\}}{|1 - s_{22}\Gamma_L|^2 - |s_{11} + \Gamma_L(s_{12}s_{21} - s_{11}s_{22})|^2} \qquad (2.42)$$

Thus, the network efficiency depends on Γ_L but not on Γ_G. When $\Gamma_L = 0$,

$$\eta_{(\Gamma_L = 0)} = \frac{|s_{21}|^2}{1 - |s_{11}|^2} \qquad (2.43)$$

On comparing eqns. 2.16 and 2.43, it is seen that the dissipative component of the attenuation A_d and the network efficiency with a matched load are related as follows:

$$A_d = 10 \log_{10} \frac{1}{\eta_{(\Gamma_L = 0)}} \qquad (2.44)$$

2.7 Available power transmission factor

The available power transmission factor τ is defined in the following manner:

$$\tau = \frac{\text{maximum available power at output of 2-port}}{\text{maximum available power from source}} \quad (2.45)$$

The power delivered to the load is given by eqn. 2.6, which readily simplifies to

$$P_2 = \frac{|e|^2 |s_{21}|^2 (1 - |\Gamma_L|^2)}{Z_0 |1 - \Gamma_G s_{11}|^2 \cdot |1 - \Gamma_2 \Gamma_L|^2} \quad (2.46)$$

where

$$\Gamma_2 = s_{22} + \frac{s_{12} \Gamma_G s_{21}}{1 - \Gamma_G s_{11}} \quad (2.47)$$

P_2 reaches its maximum value when Γ_L is the complex conjugate of Γ_2. Thus, if we let $\Gamma_2 = x' + jy'$ and $\Gamma_L = x' - jy'$, it follows from eqn. 2.46 that

$$(P_2)_{max} = \frac{|e|^2 |s_{21}|^2}{Z_0 |1 - \Gamma_G s_{11}|^2 \cdot \{1 - |\Gamma_2|^2\}} \quad (2.48)$$

From eqns. 2.23, 2.45 and 2.48, we get

$$\tau = \frac{|s_{21}|^2 \{1 - |\Gamma_G|^2\}}{|1 - \Gamma_G s_{11}|^2 \{1 - |\Gamma_2|^2\}} \quad (2.49)$$

$$= \frac{|s_{21}|^2 \{1 - |\Gamma_G|^2\}}{|1 - \Gamma_G s_{11}|^2 - |s_{22} + \Gamma_G (s_{12} s_{21} - s_{11} s_{22})|^2} \quad (2.50)$$

It is seen that τ depends on Γ_G but not on Γ_L.

2.8 Mismatch loss factor

Again referring to Fig. 2.3, the mismatch loss factor M_1 at the interface between the generator and the 2-port is defined as follows:

$$M_1 = \frac{\text{power absorbed at inlet of 2-port}}{\text{maximum available power from source}} \quad (2.51)$$

Using eqn. 2.23, it follows that

$$M_1 = \frac{\dfrac{|a_1|^2}{Z_0}\{1 - |\Gamma_1|^2\}}{\dfrac{|e|^2}{Z_0\{1 - |\Gamma_G|^2\}}} \qquad (2.52)$$

As

$$a_1 = \frac{e}{1 - \Gamma_G\Gamma_1}$$

we finally get

$$M_1 = \frac{\{1 - |\Gamma_G|^2\}\{1 - |\Gamma_1|^2\}}{|1 - \Gamma_G\Gamma_1|^2} \qquad (2.53)$$

At the interface between the 2-port and the load, the mismatch loss factor is given by

$$M_2 = \frac{\text{power absorbed by load}}{\text{maximum available power at output of 2-port}} \qquad (2.54)$$

Hence, from eqns. 2.46 and 2.48, we get

$$M_2 = \frac{\{1 - |\Gamma_2|^2\}\{1 - |\Gamma_L|^2\}}{|1 - \Gamma_2\Gamma_L|^2} \qquad (2.55)$$

Combining the expressions derived in the last three sections, it is seen that

$$\frac{\text{power absorbed by load}}{\text{maximum available power from source}} = M_1\eta = \tau M_2 \qquad (2.56)$$

2.9 Cascaded 2-ports

When any microwave system is assembled, several 2-port devices are usually connected in cascade and it is often necessary to know the s-parameters of the composite 2-port that is formed.[1,2,6,7] The necessary equations for doing this can be written down straight away by making use of signal flow graphs and Mason's non-touching loop rule (see Appendix 2).

The signal flow graph for two cascade-connected 2-ports between a matched source and a matched load is shown in Fig. 2.4. In this signal flow graph there is one first-order loop $m_{22}n_{11}$ and no higher order loops, so the s-parameters of the composite 2-port are seen at once to be

$$s_{11} = m_{11} + \frac{m_{21}n_{11}m_{12}}{1 - m_{22}n_{11}} \qquad (2.57)$$

$$s_{12} = \frac{n_{12}m_{12}}{1 - m_{22}n_{11}} \tag{2.58}$$

$$s_{21} = \frac{m_{21}n_{21}}{1 - m_{22}n_{11}} \tag{2.59}$$

$$s_{22} = n_{22} + \frac{n_{12}m_{22}n_{21}}{1 - m_{22}n_{11}} \tag{2.60}$$

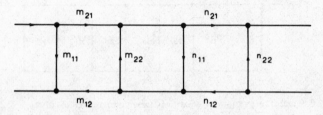

Fig. 2.4 Signal flow graph of two cascade-connected 2-ports

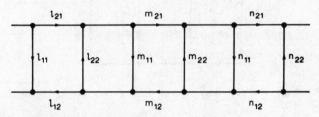

Fig. 2.5 Signal flow graph of three cascade-connected 2-ports

Fig. 2.5 gives the signal flow graph for three cascade connected 2-ports between a matched source and a matched load. We now have three first-order loops, $l_{22}m_{11}$, $m_{22}n_{11}$ and $l_{22}m_{21}n_{11}m_{12}$ and one second-order loop, $l_{22}m_{11}m_{22}n_{11}$. Thus, for the composite 2-port, we now have

$$s_{11} = l_{11} + \frac{l_{21}m_{11}l_{12}(1 - m_{22}n_{11}) + l_{21}m_{21}n_{11}m_{12}l_{12}}{1 - (l_{22}m_{11} + m_{22}n_{11} + l_{22}m_{21}n_{11}m_{12}) + l_{22}m_{11}m_{22}n_{11}} \tag{2.61}$$

$$s_{21} = \frac{l_{21}m_{21}n_{21}}{1 - (l_{22}m_{11} + m_{22}n_{11} + l_{22}m_{21}n_{11}m_{12}) + l_{22}m_{11}m_{22}n_{11}} \tag{2.62}$$

and corresponding expressions for s_{12} and s_{22}.

2.10 Measurement of the substitution loss of a variable attenuator between isolators

Some workers always keep unmatched isolators on either side of a precision variable attenuator and submit the complete assembly for calibration. It is therefore important to determine what happens in this case. The signal flow graph for this arrangement is shown in Fig. 2.6. An exact solution of this problem is straightforward but tedious

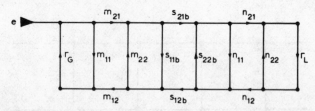

Fig. 2.6 Signal flow graph of attenuator between unmatched isolators

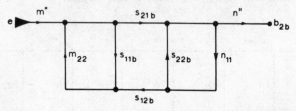

Fig. 2.7 Simplified signal flow graph of an attenuator between unmatched isolator

because the signal flow graph contains 10 first-order loops and numerous higher-order ones. However, the problem is greatly simplified if we assume that both isolators have infinite reverse attenuation and therefore let $m_{12} = n_{12} = 0$. The two end parts of the signal flow graph can now be simplified by using the loop elimination rule and we then get the signal flow graph shown in Fig. 2.7, where

$$m'' = \frac{m_{21}}{1 - \Gamma_G m_{11}} \quad \text{and} \quad n'' = \frac{n_{21}}{1 - n_{22}\Gamma_L} \quad (2.63), (2.64)$$

From the non-touching loop rule, we get

$$\frac{b_{2b}}{e} = \frac{m'' s_{21b} n''}{1 - (m_{22}s_{11b} + s_{22b}n_{11} + m_{22}s_{21b}n_{11}s_{12b}) + m_{22}s_{11b}s_{22b}n_{11}}$$

$$(2.65)$$

Suppose now that the variable attenuator is changed from its zero position to another position where its characteristics are denoted by e instead of b subscripts. In this case, we have

$$\frac{b_{2e}}{e} = \frac{m''s_{21e}n''}{1 - (m_{22}s_{11e} + s_{22e}n_{11} + m_{22}s_{21e}n_{11}s_{12e}) + m_{22}s_{11e}s_{22e}n_{11}}$$

(2.66)

The substitution loss is therefore

$$L_s = 20 \log_{10} \frac{|b_{2b}|}{|b_{2e}|}$$

(2.67)

$$= 20 \log_{10} \frac{|s_{21b}| \cdot |(1 - m_{22}s_{11e})(1 - n_{11}s_{22e}) - m_{22}n_{11}s_{12e}s_{21e}|}{|s_{21e}| \cdot |(1 - m_{22}s_{11b})(1 - n_{11}s_{22b}) - m_{22}n_{11}s_{12b}s_{21b}|}$$

(2.68)

By comparing eqn. 2.68 with eqn. 2.21, we can now see that:

(a) the value of the substitution loss obtained in this way is totally independent of the source and load reflection coefficients, Γ_G and Γ_L

(b) eqn. 2.68 can be obtained from eqn. 2.21 by simply replacing Γ_G by m_{22} and Γ_L by n_{11}

(c) the quantity that is measured in this way is not the incremental attenuation.

If the isolators are not perfect, the substitution loss becomes slightly dependent on the values of Γ_G and Γ_L.

2.11 Mismatch error and uncertainty

If an attenuation measurement is made without the source and load being perfectly matched there will be an error in the result. This error, which occurs because insertion loss is what is actually measured, is known as the mismatch error[2,8-12] and it is given in decibels by

$$M_I = L_I - A$$

(2.69)

Thus, the mismatch error is simply the difference between the insertion loss and the attenuation. Hence, using eqns. 2.9 and 2.10 we find that

$$M_I = 20 \log_{10} \frac{|(1 - \Gamma_G s_{11})(1 - \Gamma_L s_{22}) - \Gamma_G \Gamma_L s_{12}s_{21}|}{|1 - \Gamma_G \Gamma_L|}$$

(2.70)

All of the independent variables in this last equation are complex quantities and all phase relationships are possible. If both the real and the imaginary parts of all of these quantities are known, M_I can be found precisely. However, if only their magnitudes are known, as is often the case, then the mismatch uncertainty is given by

$$(M_I)_{limit} = 20 \log_{10}$$

$$\frac{1 \pm \{|\Gamma_G s_{11}| + |\Gamma_L s_{22}| + |\Gamma_G \Gamma_L s_{11} s_{22}| + |\Gamma_G \Gamma_L s_{12} s_{21}|\}}{1 \mp |\Gamma_G \Gamma_L|} \quad (2.71)$$

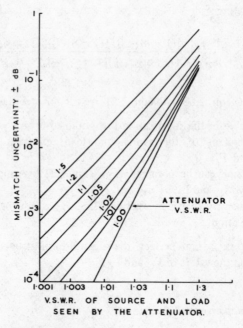

Fig. 2.8 Mismatch uncertainty
For a symmetrical attenuator giving a loss > 20 dB

Fig. 2.8 shows a family of mismatch uncertainty curves calculated from eqn. 2.71. These curves are expressed in terms of voltage standing wave ratios, rather than reflection coefficients, to satisfy a preference shown by most microwave engineers. If the attenuator being measured is symmetrical and has a v.s.w.r. of 1·05, the source and load v.s.w.r.s must be less than 1·005 to keep the mismatch uncertainty below 0·001 dB. The importance of this mismatch uncertainty cannot be over emphasised. When very precise attenuation measurements are carried out in a standards laboratory, the mismatch uncertainty is frequently the largest term contributing to the systematic uncertainty. Practical techniques that are used to minimise or make corrections for the mismatch error are discussed in Chapters 5 and 11.

When a variable attenuator is calibrated relative to its 'zero' or 'datum' position, each complete measurement yields a value for the substitution loss (see Section 2.3). Thus, in this case, the mismatch error is given by

$$M_s = L_s - A_{inc} \qquad (2.72)$$

or, stated in words, it is simply the difference between the substitution loss and the incremental attenuation. Combining eqns. 2.21, 2.22 and 2.72 we get

$$M_s = 20 \log_{10} \frac{|(1 - \Gamma_G s_{11e})(1 - \Gamma_L s_{22e}) - \Gamma_G \Gamma_L s_{12e} s_{21e}|}{|(1 - \Gamma_G s_{11b})(1 - \Gamma_L s_{22b}) - \Gamma_G \Gamma_L s_{12b} s_{21b}|}$$

$$(2.73)$$

If only the magnitudes of the quantities on the right-hand side of eqn. 2.73 are known, the mismatch uncertainty is

$$(M_s)_{limit} = 20 \log_{10} \qquad (2.74)$$

$$\frac{1 \pm \{|\Gamma_G s_{11e}| + |\Gamma_L s_{22e}| + |\Gamma_G \Gamma_L s_{11e} s_{22e}| + |\Gamma_G \Gamma_L s_{12e} s_{21e}|\}}{1 \mp \{|\Gamma_G s_{11b}| + |\Gamma_L s_{22b}| + |\Gamma_G \Gamma_L s_{11b} s_{22b}| + |\Gamma_G \Gamma_L s_{12b} s_{21b}|\}}$$

With similar values for the s-parameters, $(M_s)_{limit}$ is roughly twice as large as $(M_I)_{limit}$.

When Γ_G, Γ_L, s_{11b}, s_{22b}, s_{11e} and s_{22e} are all very small, and the attenuator being measured is reciprocal so that $s_{12b} = s_{21b}$ and $s_{12e} = s_{21e}$, eqn. 2.73 can be simplified to

$$M_s \approx 20 \log_{10} |1 - \Gamma_G(s_{11e} - s_{11b}) - \Gamma_L(s_{22e} - s_{22b})$$

$$+ \Gamma_G \Gamma_L(s_{21b}^2 - s_{21e}^2)| \qquad (2.75)$$

If the input and output reflection coefficients of the unknown stay constant as the attenuation is varied, the second and third terms in the modulus vanish and, for small changes of attenuation, the mismatch error is then extremely small.[13]

2.12 Leakage

Another source of error in any attenuation measurement is leakage. Let A_A denote the attenuation in dB through the attenuator under test, and let A_L denote the attenuation in dB through a leakage path shunting this attenuator. Furthermore, let E_G represent the wave

amplitude at the input to both the attenuator and the leakage path, and
let the waves arriving at the receiver through the attenuator and the
leakage path have amplitudes E_A and E_L, respectively, and a phase
difference ϕ between them. Then, using the cosine rule, the resultant
amplitude at the receiver is found to be

$$E_R = \{E_A^2 + E_L^2 + 2E_A E_L \cos \phi\}^{1/2} \tag{2.76}$$

and the error in dB in the attenuation measurement due to leakage is
given by

$$\delta A_L = 20 \log_{10} \frac{E_G}{E_R} - 20 \log_{10} \frac{E_G}{E_A} \tag{2.77}$$

$$= 20 \log_{10} \left\{ 1 + \left(\frac{E_L}{E_A}\right)^2 + 2 \frac{E_L}{E_A} \cos \phi \right\}^{-1/2} \tag{2.78}$$

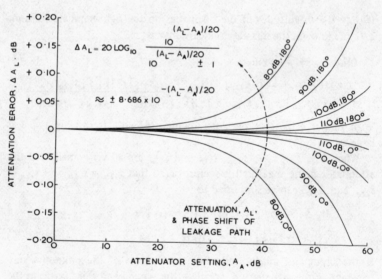

Fig. 2.9 Error caused by external leakage

The upper and lower limits of this leakage error are obtained with
$\cos \phi = \pm 1$ and are given by

$$\Delta A_L = 20 \log_{10} \left(1 \pm \frac{E_L}{E_A} \right)^{-1} \tag{2.79}$$

However,

$$A_A = 20 \log_{10} \frac{E_G}{E_A} \quad \text{and} \quad A_L = 20 \log_{10} \frac{E_G}{E_L} \qquad \begin{matrix} (2.80), \\ (2.81) \end{matrix}$$

Thus
$$\frac{E_L}{E_A} = 10^{-(A_L-A_A)/20} \tag{2.82}$$

Substituting eqn. 2.82 in eqn. 2.79 we finally get

$$\Delta A_L = 20 \log_{10} \{1 \pm 10^{-(A_L-A_A)/20}\}^{-1} \tag{2.83}$$

When $A_L \gg A_A$

$$\Delta A_L \approx \mp 8 \cdot 686 \{10^{-(A_L-A_A)/20}\} \tag{2.84}$$

A general family of leakage error curves is shown in Fig. 2.9. To keep the leakage error below $0 \cdot 001$ dB, A_L must be 80 dB greater than A_A. Thus, great care must be taken to eliminate all traces of leakage from a precision attenuation measuring system.

2.13 References

1 KERNS, D.M., and BEATTY, R.W.: 'Basic theory of waveguide junctions and introductory microwave network analysis' (Pergamon, 1967)

2 BEATTY, R.W.: 'Applications of waveguide and circuit theory to the development of accurate microwave measurement methods and standards'. NBS Monograph 137, August 1973

3 MILLER, C.K.S., DAYWITT, W.C., and ARTHUR, M.G.: 'Noise standards, measurements and receiver noise definitions', *Proc. IEEE*, 1967, **55** pp. 865–877

4 OTOSHI, T.Y.: 'The effect of mismatched components on microwave noise-temperature calibrations', *IEEE Trans.*, 1968, **MTT-16**, pp. 675–686

5 BLUNDELL, D.J., HOUGHTON, E.W., and SINCLAIR, M.W.: 'Microwave noise standards in the United Kingdom', *IEEE Trans.*, 1972, **IM-21**, pp. 484–488

6 BEATTY, R.W.: 'Cascade-connected attenuators', *J. Res. NBS*, 1950, **45**, pp. 231–235

7 BEATTY, R.W.: 'Cascade-connected attenuators', *Proc. IRE*, 1950, **38**, p. 1190

8 BEATTY, R.W.: 'Mismatch errors in the measurement of ultrahigh-frequency and microwave variable attenuators', *J. Res. NBS*, 1954, **52**, pp. 7–9

9 SCHAFER, G.E., and RUMFELT, A.Y.: 'Mismatch errors in cascade connected variable attenuators', *IRE Trans.*, 1959, **MTT-7**, pp. 447–453

10 LEED, D.: 'Use of flow graphs to evaluate mistermination errors in loss and phase measurements', *IRE Trans.*, 1961, **MTT-9**, p. 454

11 'Microwave mismatch error analysis'. Hewlett Packard Application Note 56, October 1967

12 BEATTY, R.W.: 'Discussion of effect of realisability conditions on estimated limits of mismatch error in the calibration of fixed attenuators, *IEEE Trans.*, 1968, **MTT-16**, p. 976

13 ENGEN, G.F., and BEATTY, R.W.: 'Microwave attenuation measurements with accuracies from $0 \cdot 0001$ to $0 \cdot 06$ decibel over a range of $0 \cdot 01$ to 50 decibels', *J. Res. NBS*, 1960, **64C**, pp. 139–145

Chapter 3

D.C., a.f. and i.f. attenuation standards

In several of the microwave attenuation measuring systems that are described later in this book, the attenuation reference standard operates at d.c., a.f. or i.f. and is usually a Kelvin-Varley voltage divider, an inductive voltage divider or an intermediate frequency piston attenuator. These three devices are described in this chapter and particular attention is given to the sources of error associated with them.

3.1 Kelvin-Varley voltage dividers

The Kelvin-Varley voltage divider[1-6] contains only resistors and switches and operates right down to zero frequency. Fig. 3.1 shows a four decade divider of this type, which enables the ratio of the output voltage to the input voltage to be varied from 0 to 1 in steps of 1 part in 10^4. Eleven equal resistors are used in each decade except the last one which contains ten. The switches in each decade, except the last, always span two resistors, and the values of the resistors in each succeeding decade become smaller by a factor of five. A brief study of Fig. 3.1 will show that this arrangement, when unloaded, gives exact decimal switching of the output voltage relative to the input voltage. The total resistance of the last decade is $10R$. Two adjacent resistors in the penultimate decade are shunted by the last decade at each dial setting, so, in effect, this penultimate decade contains 10 resistors, each of value $5R$, in series with each other, and so on.

The arrangement used in the last stage of a Kelvin-Varley voltage divider was patented by Professor W. Thomson (Lord Kelvin) and Fleeming Jenkin in 1860, and the ingenious way of connecting several decades together, so that exact decimal switching is obtained, was devised by C.F. Varley in 1866.[1]

When a Kelvin-Varley voltage divider is unloaded, its input resistance R_{in} stays constant as the switch settings are varied. For the divider shown in Fig. 3.1, $R_{in} = 1250\,R$. However, the output resistance does vary with the switch settings and it has a maximum value slightly greater than $R_{in}/4$ when the source resistance is zero.

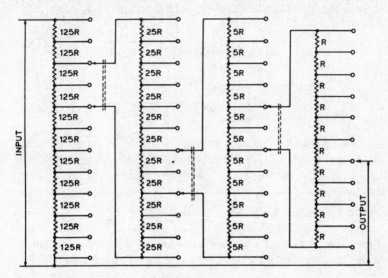

Fig. 3.1 4-decade Kelvin-Varley voltage divider

The frequency response of a Kelvin-Varley divider is similar to that of a simple RC low-pass filter. When good constructional techniques are used, the 3 dB bandwidth is in the region of 100 kHz when $R_{in} = 100\,k\Omega$, and 10 MHz when $R_{in} = 1\,k\Omega$.

The power dissipated in each of the bridged resistors in any particular decade is only a quarter of that dissipated in each of the other 9 resistors. This unequal heating necessitates the use of very low temperature coefficient resistors.

Although decimal switching is normally used, the Kelvin-Varley principle can be readily applied to a switching system with any numerical base N_b. We then require N_b resistors in the last stage, $(N_b + 1)$ resistors in each of the other stages, and the values of the resistors in each succeeding decade must be smaller by a factor $N_b/2$. Thus, in a binary Kelvin-Varley voltage divider, $N_b = 2$ and all resistors have the same value. A voltage divider of this type has been described by Neff.[6]

In a Kelvin-Varley voltage divider, errors are caused by departures from the nominal resistance values and by resistance in the switch contacts and connecting leads. For a quick insight to this matter, let

us consider the last two stages of a binary Kelvin-Varley voltage divider
(see Fig. 3.2). Deviations from the nominal resistance values are denoted by $\Delta_1, \Delta_2, \Delta_3$, etc. With the switches set as shown in the diagram
to give a ratio of 1/4, it follows at once from Kirchhoff's laws that

$$\left(\frac{V_{out}}{V_{in}}\right)_{1/4} = \frac{R_r}{\{R(1+\Delta_1)+R_r\}} \cdot \frac{R(1+\Delta_5)}{\{R(1+\Delta_4)+R(1+\Delta_5)\}} \quad (3.1)$$

where

$$R_r = \frac{\{R(1+\Delta_2)+R(1+\Delta_3)\}\{R(1+\Delta_4)+R(1+\Delta_5)\}}{R(1+\Delta_2)+R(1+\Delta_3)+R(1+\Delta_4)+R(1+\Delta_5)} \quad (3.2)$$

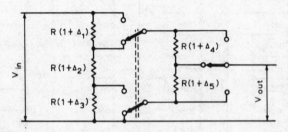

Fig. 3.2 2-stage binary Kelvin-Varley voltage divider

From eqns. 3.1 and 3.2 we find, after making certain approximations
which are fully justified when the Δ terms are very small, that

$$\left(\frac{V_{out}}{V_{in}}\right)_{1/4} = \frac{1}{4} - \frac{\Delta_1}{8} + \frac{\Delta_2 + \Delta_3 - 3\Delta_4 + 5\Delta_5}{32} \quad (3.3)$$

A ratio of $\frac{1}{2}$ can be obtained either with the first switch down and the
second switch at the top, or with the first switch up and the second one
at the bottom. Similar treatments to that given above then yield,
respectively,

$$\left(\frac{V_{out}}{V_{in}}\right)'_{1/2} = \frac{1}{2} - \frac{\Delta_1}{4} + \frac{\Delta_2 + \Delta_3 + \Delta_4 + \Delta_5}{16} \quad (3.4)$$

and

$$\left(\frac{V_{out}}{V_{in}}\right)''_{1/2} = \frac{1}{2} + \frac{\Delta_3}{4} - \frac{\Delta_1 + \Delta_2 + \Delta_4 + \Delta_5}{16} \quad (3.5)$$

Finally, with the switches set to give a ratio of $\frac{3}{4}$, we find that

$$\left(\frac{V_{out}}{V_{in}}\right)_{3/4} = \frac{3}{4} - \frac{\Delta_1}{32} - \frac{\Delta_2}{32} + \frac{\Delta_3}{8} - \frac{5\Delta_4}{32} + \frac{3\Delta_5}{32} \quad (3.6)$$

When $\Delta_1 = \Delta_2 = \Delta_3 = \Delta_4 = \Delta_5$, it is seen from eqns. 3.3 to 3.6 that
there are no errors in any of the ratios. When all of the resistor values

are in error by either $+\Delta$ or $-\Delta$ and we have the worst possible sign combinations, the maximum errors in the ratios are seen to be $\pm\frac{7}{16}\Delta$ at the $\frac{1}{4}$ and $\frac{3}{4}$ settings and $\pm\frac{1}{2}\Delta$ at the $\frac{1}{2}$ setting. Thus, if a divider of this type is made with $\pm 0\cdot 01\%$ resistors, the maximum error in the output voltage will be $\pm 0\cdot 005\%$ of the input. In this very simple circuit, it is seen from eqns. 3.3 to 3.6 that the most critical resistors are the top and bottom ones in the first stage.

This type of treatment can be readily extended to more and more complex voltage dividers, but the mathematical expressions become very lengthy. A complete analysis of a four decade Kelvin-Varley divider has been given by Dunn.[5] In a multistage divider, extremely tight tolerances are required on the resistors in the first stage, but the requirements become less and less stringent as one moves towards the output stage.

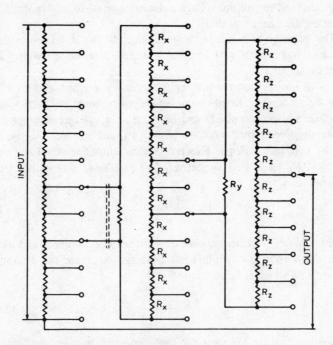

Fig. 3.3 Shunting technique that avoids the use of inconveniently large or small resistor values in a Kelvin-Varley divider

The circuit shown in Fig. 3.1 can sometimes require inconveniently large or small resistance values. This difficulty can be overcome by using the shunting technique[2] shown in Fig. 3.3. For correct decimal switching, the requirement is

$$2R_x = \frac{R_y \times 10R_z}{R_y + 10R_z} \tag{3.7}$$

Rearranging, we get

$$R_y = \frac{10R_x R_z}{5R_z - R_x} \tag{3.8}$$

Thus, this shunting technique can be used whenever $R_x < 5R_z$. If $100\,\Omega$ resistors are used throughout the penultimate and final stages (i.e. $R_x = R_z = 100\,\Omega$), the required value for the shunt resistance R_y is found from eqn. 3.8 to be $250\,\Omega$.

Due to unavoidable resistance in the leads and switches of a Kelvin-Varley divider, there is a small output voltage when all of the dials are set at zero and the output voltage is not quite equal to the input voltage when the dials are set at unity.

The linearity of a Kelvin-Varley voltage divider is defined in three different ways, namely, 4 terminal linearity, 3 terminal linearity and absolute linearity.

Let us assume that the low input terminal is earthed, and let a voltage V_{in} be applied between the high input terminal and earth. When the dials are set at zero, D and unity, let the voltages developed between the two output terminals be V_Z, V_D and V_1, respectively, and let the voltages developed between the low output terminal and earth be V_{LZ}, V_{LD} and V_{L1}, respectively (see Fig. 3.4a). The four terminal linearity

$$\zeta_4 = \frac{V_D - DV_{in}}{V_{in}} \tag{3.9}$$

If the output is taken between the output high terminal and earth, with the output low terminal left floating, we obtain the 3 terminal linearity ζ_3, which is defined as follows:

$$\zeta_3 = \frac{V_D + V_{LD} - DV_{in}}{V_{in}} \tag{3.10}$$

Absolute linearity is a more difficult concept. On a graph of voltage between the output terminals versus the divider setting D, let a straight line be drawn between the points $(0, V_Z)$ and $(1, V_1)$ as shown in Fig. 3.4b. The equation for this straight line is

$$V_{D,abs} = (V_1 - V_Z)D + V_Z \tag{3.11}$$

Absolute linearity ζ_{abs}, is defined as follows:

$$\zeta_{abs} = \frac{V_D - V_{D,abs}}{V_{in}} \qquad (3.12)$$

Combining eqns. 3.11 and 3.12, we get

$$\zeta_{abs} = \frac{V_D - V_Z - D(V_1 - V_Z)}{V_{in}} \qquad (3.13)$$

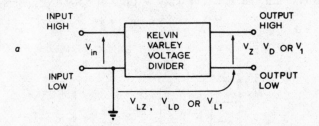

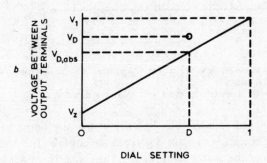

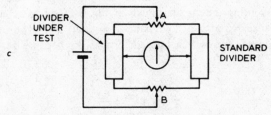

Fig. 3.4 Diagrams related to the three definitions of linearity
  a Circuit for determining the three definitions of linearity
  b Graph for defining absolute linearity
  c Circuit for comparing the absolute linearities of two Kelvin-Varley
    dividers

When $D = 0$, $V_D = V_Z$ and it follows from eqns. 3.9, 3.10 and 3.13
that

$$\zeta_4 = \frac{V_Z}{V_{in}} \qquad \zeta_3 = \frac{V_Z + V_{LZ}}{V_{in}} \qquad \zeta_{abs} = 0 \,.$$

When $D = 1$, $V_D = V_1$ and it is seen that

$$\zeta_4 = \frac{V_1 - V_{in}}{V_{in}} \qquad \zeta_3 = \frac{V_1 + V_{L1} - V_{in}}{V_{in}} \qquad \zeta_{abs} = 0$$

The absolute linearities of two Kelvin-Varley dividers can be compared by using the circuit shown in Fig. 3.4c. Potentiometers A and B are adjusted until the zero and unity settings of both dividers coincide.

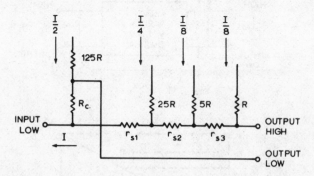

Fig. 3.5 Diagram showing how a resistor R_c can be used to improve the four terminal linearity
Resistances in switch contacts and leads are denoted by r_{s1}, r_{s2} and r_{s3}

To achieve good four-terminal linearity at the very bottom end of the scale, it is necessary to make V_Z as low as possible. This voltage is developed across the switch contacts and the connecting leads by the currents that flow through the various decades, and it can be cancelled out by using the circuit shown in Fig. 3.5. An extra compensating resistance R_c of very low value is inserted in the bottom end of the first decade and, with a 4-decade divider, V_Z becomes zero when

$$\frac{I}{2}R_c = \frac{I}{2}r_{s1} + \frac{I}{4}r_{s2} + \frac{I}{8}r_{s3} \qquad (3.14)$$

i.e. when

$$R_c = r_{s1} + \frac{r_{s2}}{2} + \frac{r_{s3}}{4} \qquad (3.15)$$

Several different techniques for calibrating Kelvin-Varley voltage dividers have been developed and these have been described by Morgan and Riley,[3] Fryer[4] and Dunn.[5]

In recent years, great progress has been made in the wire-wound resistor field and amazingly accurate ultra-stable resistors can now be produced. As a result of this progress, Kelvin-Varley voltage dividers can now be obtained, commercially, with an absolute linearity of ± 1 part in 10^7 and a long term stability of ± 1 part in 10^6 per year.

3.2 Inductive voltage dividers

3.2.1 Introduction
The inductive voltage divider (IVD), sometimes called a ratio trans-former, forms an exceptionally accurate variable attenuation standard at audio frequencies. Fig. 3.6 shows the circuit diagram of a 7 decade IVD. It consists of 7 very accurately tapped autotransformers which are connected together with high class switches in such a way that the ratio of the output voltage to the input voltage can be varied from 0 to 1, in steps of 1 part in 10^7. Precision IVDs can be designed to operate at any frequency between about 10 Hz and 100 kHz and greatest accuracy is obtained at a frequency of about 1 kHz.

The IVD can be traced back to an invention by A.D. Blumlein.[7] In 1928, he proposed replacement of the two resistive ratio arms of an a.c. bridge by a pair of tightly coupled inductors. This invention brought the three terminals of the ratio arms to almost the same potential and eliminated the need for a Wagner earth.[8] Since 1928, many different bridges based on this principle have been described in the literature.[9-15]

Attention was turned next to multi-decade inductive voltage divi-ders[16-28] and, in 1962, Hill and Miller[17] described a 7-decade one, similar to that in Fig. 3.6, with an accuracy of better than 1 part in 10^7. During the next decade even further progress was made, and, in 1972, Hill[28] described a 3-decade low-frequency IVD in which the errors of voltage division do not exceed 5 parts in 10^9 of the input. Eight-decade IVDs with an accuracy of 4 parts in 10^8 of the input are now commercially available. In recent years, IVDs have found numerous applications in many different branches of electronics. Their vital role in microwave attenuation measuring equipment will be described in Chapter 8.

3.2.2 Construction of inductive voltage dividers
Many different forms of construction have been investigated. Good results can be achieved by taking 10 exactly equal lengths of enamel-covered copper wire from the same reel and twisting them together (about 50 twists/m) in either a uniform or random manner. The re-sulting 'rope' is arranged to form a single layer around a very high permeability toroidal core.[17] Even better results can be achieved by arranging the 10-strand cable so that it forms a double layer on the inner circumference and a single layer on the outer circumference.[23]

Nine appropriate pairs of wire ends from the 10 strand cable are then joined together so that 10 series-aiding coils are obtained.

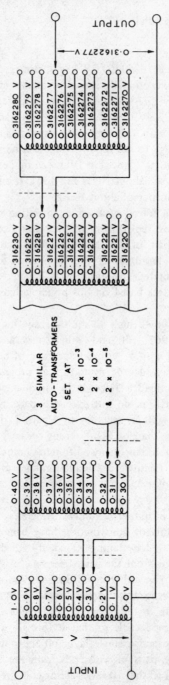

Fig. 3.6 Simplified circuit diagram of a 7-decade inductive voltage divider
Set to give a voltage ratio of 0·3162277 = 10 dB

In many IVD applications, switching transients are highly undersirable and these can be eliminated by using a switching technique that never allows either short circuits or open circuits to occur. One switch circuit that meets these requirements has been described by Hill and Miller[17] and this is shown in Fig. 3.7. For decimal switching, a nineteen-way double-pole make-before-break switch is needed and $1 \, k\Omega$ has been found to be a suitable value for the resistors.

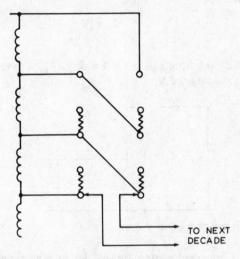

Fig. 3.7 Switching arrangement that minimises transients
2-pole 19-way make before break switch is used

The great success achieved with IVDs is closely related to the discovery of a remarkable magnetic material at the Bell Telephone Laboratories in 1942.[29] Known as supermalloy, it is composed of 79% nickel, 15% iron, 5% molybdenum and 1% of other elements. It is heat treated in pure hydrogen at $1300°C$ and then cooled at a critical rate. Spirally-wound supermalloy tape cores have an initial permeability in the range 60 000 to 150 000, a maximum permeability in the range 200 000 to 400 000 and a hysteresis loss of less than $1 \, J/m^3/Hz$ at $0·5$ T. It will be shown later that very high permeability and extremely low hysteresis loss are vital requirements for cores used in IVDs and reference to the magnetic data tables given by Bardell[30] will show that supermalloy is quite unique in both of these respects. Thus, it is very widely used in precision IVDs. Supermalloy is sold under various trade names such as 'Permalloy Super C' and 'Supermumetal'. Supermalloy can be rolled down to about $5 \, \mu m$ and it can be insulated with a $1 \, \mu m$ thick film of magnesia.[29] Supermalloy cores for IVDs are wound like a clock spring and are protected by a casing of nylon or some similar material. The casing is often packed with silicone grease to reduce the acoustic noise from the core.

3.2.3 Basic theory related to inductive voltage dividers

First of all, let us consider the unloaded binary inductive voltage divider shown in Fig. 3.8. Then, from elementary mutual inductance theory, it follows that

$$\frac{V_{out}}{V_{in}} = \frac{L_2 + M}{L_1 + L_2 + 2M} \tag{3.16}$$

If $L_1 = L_2$, it is seen from eqn. 3.16 that V_{out}/V_{in} is exactly equal to $\frac{1}{2}$ regardless of the value of M.

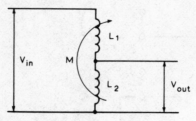

Fig. 3.8 Simplest inductive voltage divider

Taking this argument a step further, let us now suppose that an unloaded autotransformer with 10 sections is tapped n sections above the earthed end. Let L_S denote the self inductance of each section, and let M now denote the mutual inductance between any pair of sections. Then, as the same current flows through all ten sections, the coupled inductance of each section is $L_S + 9M$, and so we now get

$$\frac{V_{out}}{V_{in}} = \frac{n(L_S + 9M)}{10(L_S + 9M)} = \frac{n}{10} \tag{3.17}$$

Again, we see that exact voltage division is obtained regardless of the value of M. Thus, for perfect division, the requirement is not 100% coupling but exact equality of the 10 self inductances and also exact equality of the 45 mutual inductances.

So far, we have ignored the hysteresis and eddy current losses and the resistances of the windings. To enable us to see the overall picture quickly, the basic equations for a toroidal coil will now be set down.

Let us consider a toroidal core of effective relative permeability μ_t with a rectangular cross-section of area A_c. Let its width be b, its mean circumference be c, and let its inner and outer radii be denoted by a_i and a_0, respectively. Then

$$A_c = b(a_0 - a_i) \quad \text{and} \quad c = \pi(a_i + a_0)$$

Let N turns of wire of diameter d and resistivity ρ_{cu} be wound on this core. Then, in SI units, it is found from standard textbooks[31-33] that the total inductance

$$L_t = \frac{\mu_t \mu_0 N^2 b}{2\pi} \log_e \frac{a_0}{a_i} \tag{3.18}$$

$$\approx \frac{\mu_t \mu_0 N^2 A_c}{c} \tag{3.19}$$

the shunt resistance across the entire winding due to hysteresis losses is given by

$$R_h = \frac{2\pi^2 f B^2 N^2 A_c}{\xi c} \tag{3.20}$$

the shunt resistance across the entire winding due to eddy current losses is given by

$$R_e = \frac{12 N^2 A_c \rho_s}{t^2 c} \tag{3.21}$$

the total series resistance

$$r_t = \frac{8 K \rho_{cu} N (a_0 - a_i + b)}{\pi d^2} \tag{3.22}$$

and the maximum r.m.s. voltage that can be applied across the entire winding is given by

$$V_{max} = \sqrt{2}\pi B_{max} A_c N f \tag{3.23}$$

where B is the flux density, B_{max} is the maximum permissible value for B, f is the frequency, t is the thickness of the core material, ξ is the hysteresis loss/m^3/Hz at flux density B, $\mu_0 = 4\pi/10^7$ H/m, and ρ_s is the specific resistance of the core material. In eqn. 3.22, K is a constant, lying between 1 and 2, which accounts for the increase in the total wire length due to its finite diameter, the insulation on it, the casing around the core and the twisting referred to earlier.

When $a_0 = 4a_i/3$, eqn. 3.19 yields a value for L_t which is 0·68% lower than the value given by eqn. 3.18.

It has been found that the mean leakage inductance per section $\bar{l}$ in an IVD is directly proportional to the length of the 10-strand cable and is independent of the number of turns around the core and the permeability of the core. A typical value[27] for the leakage inductance is $0·4\,\mu$H/m.

It will be seen later that for maximum input impedance, minimum output impedance and maximum accuracy, we want L_t, R_h and R_e to be as high as possible, and r_t and $\bar{l}$ to be as low as possible. We will now examine various steps that can be taken to achieve these objectives.

(*a*) **Wire size:** Eqn. 3.22 shows that the wire diameter d should be made as large as possible. However, if too heavy a gauge is used, the 10 strand cable will not be flexible enough to fit closely around the core and the optimum wire size appears to be 24 s.w.g. If this gauge does not provide a sufficiently low output impedance, each length of wire can be replaced by two lengths of 24 s.w.g. wire which are connected in parallel. This technique halves both the series resistance and the leakage inductance and still gives the necessary flexibility.

(*b*) **Core material:** to keep the leakage inductance and series resistance very low, the 10-strand cable should be kept as short as possible. On the other hand, a high value of inductance is required, so it follows from eqn. 3.18 or 3.19 that the permeability μ_t, should be made as high as possible. We also want ξ to be very low and ρ_s to be very high. As mentioned earlier, supermalloy is the best available material for meeting all of these requirements. Eqn. 3.21 shows that the eddy current shunt resistance is inversely proportional to the square of the lamination thickness, so very thin supermalloy tape should be used. However, a compromise is again necessary. If tape thinner than 0·025 mm is used, the space factor (i.e. the fraction of the cross-sectional area occupied by metal) falls off rapidly due to the insulating layer on the tape and imperfect contact throughout the spiral. Thus, supermalloy tape with a thickness of either 0·05 mm or 0·025 mm is normally used.

(*c*) **Core size and shape:** from eqns. 3.19, 3.20 and 3.21 it is seen that A_c/c should be made large as L_t, R_h and R_e are all directly proportional to this ratio. As mentioned earlier, the best winding configuration requires a single layer of 10-strand cable around the outer circumference and a double layer of 10-strand cable around the inner circumference.

Let d'' denote the overall diameter of the 10-strand cable, and let w represent the thickness of the casing around the core. Then, for exact packing around the outer circumference, it follows from Fig. 3.9 that

$$a_0 = \frac{Nd''}{20\pi} - w - \frac{d''}{2} \tag{3.24}$$

The inner radius depends on the exact manner in which the double layer is arranged and it can be readily calculated for any given case.

When a_i and a_0 have been chosen, the only parameter left undetermined is the core width b. When $b = a_0 - a_i$ the core has a square cross-section. Increasing b raises L_t, R_h, R_e and r_t, However, from eqns. 3.18 and 3.22, it is seen that

$$\frac{L_t}{r_t} \propto \frac{b}{a_0 - a_i + b} \qquad (3.25)$$

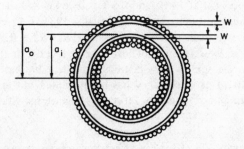

Fig. 3.9 Best winding arrangement for an IVD
Each small circle denotes the cross-section of a 10-strand cable

Hence, at the expense of an increase in the output impedance, the L_t/r_t ratio can be increased by making b greater than $(a_0 - a_i)$. If a core of the desired width cannot be found in manufacturers' catalogues, two or more standard sized cores can be placed side by side and taped together.

Typical values

To obtain some typical values for the error analysis in the next section, an example will now be given. Let us suppose that 50 turns of 10-strand cable (i.e. 500 turns altogether) are wound, in the optimum manner discussed earlier, around a supermalloy spiral tape core with an inner radius of 15 mm, an outer radius of 20 mm and a width of 20 mm. A core of this size has an adequate inner radius to accommodate a double layer on the inside and the outer radius is about 2 mm greater than the value given by eqn. 3.24. For the supermalloy core, let us assume that $\mu_t = 10^5$, $B_{max} = 0.5$ T, $\xi = 1$ J/m^3/Hz at 0.5 T, $\rho_s = 6 \times 10^{-7}$ Ωm and $t = 5 \times 10^{-5}$ m. Let 24 s.w.g. copper wire be used: then $d = 5.59 \times 10^{-4}$ m, $\rho_{cu} = 1.72 \times 10^{-8}$ Ωm and we will assume that $K = 1.6$. At a frequency of 1 kHz, we now find from the equations given earlier that $L_t = 28.7$ H, $R_h = 1.12$ MΩ, $R_e = 655$ kΩ, $r_t = 2.8$ Ω, and $V_{max} = 111$ volts r.m.s. The length of the 10-strand cable = 4 m and the leakage inductance per section ≈ 1.6 μH.

Eqns. 3.20 and 3.21 are very useful for showing the ways in which various parameters affect the values of the shunt resistances caused by hysteresis and eddy current losses. However, they do not give accurate answers, so, in practice, the value of the combined shunt resistance due to these two effects must be obtained from manufacturers data sheets. At 1 kHz and 0·5 T, a typical value given for the total power loss in a core made of 0·05 mm thick supermalloy is 5800 W/m³. The core considered earlier has a volume of $1·1 \times 10^{-5}$ m³, so the power lost in it at 1 kHz and 0·5 T is 0·064 W and the value of the combined shunt resistance is $(111)^2/0·064 = 193$ kΩ. From the results given earlier it is seen that the value of $R_e R_h/(R_e + R_h)$ is 413 kΩ. Thus, there is a discrepancy of just over 2:1 between the value obtained from the text book equations and the value based on actual measurements. This discrepancy has been pointed out and explained by both Hammond[34] and the Electrical Engineering Staff of MIT.[35]

Skin effect

The well known skin effect can affect the performance of IVDs in two different ways. Welsby[33] has shown that the critical frequency at which the skin effect starts to become important in the wire is given by

$$f_{crit} = \frac{\rho_{cu} \times 10^7}{\pi^2 d^2} \tag{3.26}$$

and, with 24 s.w.g. copper wire, this critical frequency is 55·8 kHz. Thus, skin effect in the wire is only important in high frequency IVDs. Near this critical frequency, the series resistance starts to rise above the value given by eqn. 3.22 and it soon becomes proportional to $\sqrt{f}$. The ratio of the a.c. resistance to the d.c. resistance at any frequency can be found rapidly from a table given by Terman.[36] The full theory of the skin effect in a circular conductor is missing from most of the standard works but it is given by McLachlan.[37]

The skin effect has a much more serious effect in the spiral tape core.[33] The eddy currents produce a magnetic flux which opposes the main flux in the interior of the tape and enhances it near the surface. The critical frequency at which this magnetic skin effect starts to become important is given by

$$f''_{crit} = \frac{\rho_s \times 10^7}{\mu_t \pi^2 t^2} \tag{3.27}$$

With 0·05 mm thick supermalloy tape, this critical frequency is in the region of 2·4 kHz. Above f''_{crit}, Welsby[33] has shown that L_t becomes proportional to $1/\sqrt{f}$, R_e becomes proportional to $\sqrt{f}$ and the Q becomes unity.

3.2.4 Errors in inductive voltage dividers

When the dials of a *perfect* IVD indicate a ratio D and the input voltage is denoted by V_{in}, the output voltage is $V_{in}D$. The error ζ in a practical IVD is referred to the input and is defined as follows:

$$\zeta = \frac{V_{out} - V_{in}D}{V_{in}} \tag{3.28}$$

where V_{out} is the actual output voltage at a setting D. In an IVD, errors are caused by:

(a) inequalities in the series resistances of the 10 sections in each transformer
(b) inequalities in the leakage inductances
(c) inhomogeneities in the magnetic cores
(d) distributed admittances between the windings
(e) internal loading due to connection of the later decades
(f) impedances of the connecting leads and switch contacts
(g) variations in the input voltage, frequency and ambient temperature.

All of these errors will now be discussed.

Error caused by inequalities in the windings

The low frequency equivalent circuit of a single-stage decade inductive voltage divider is shown in Fig. 3.10. In terms of the symbols used earlier

$$L = L_S + 9M = L_t/10 \qquad R_p = R_e R_h / \{10(R_e + R_h)\}$$

and

$$\bar{r} = r_t/10$$

The mean leakage inductance of each of the 10 tapped sections is denoted by $\bar{l}$. The inequalities in the series resistances and leakage inductances are accounted for by including the δr_m and δl_m terms and clearly:

$$\sum_{m=0.1}^{m=1} \delta r_m = 0 \quad \text{and} \quad \sum_{m=0.1}^{m=1} \delta l_m = 0 \qquad (3.29), (3.30)$$

Using simple a.c. theory, it is seen that

$$V_m = V_{in} \frac{\dfrac{j\omega L R_p}{R_p + j\omega L} + \bar{r} + \delta r_m + j\omega(\bar{l} + \delta l_m)}{10\left\{\dfrac{j\omega L R_p}{R_p + j\omega L} + \bar{r} + j\omega\bar{l}\right\}} \tag{3.31}$$

$$= \frac{V_{in}}{10} \left\{ 1 + \frac{\delta r_m + j\omega\delta l_m}{\dfrac{j\omega L R_p}{R_p + j\omega L} + \bar{r} + j\omega\bar{l}} \right\} \quad (3.32)$$

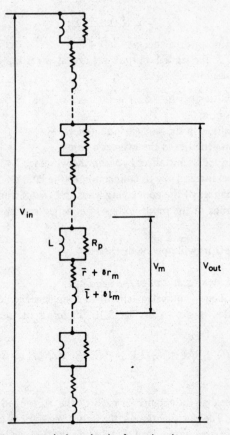

Fig. 3.10 Low-frequency equivalent circuit of one decade

Since $\bar{r} + j\omega\bar{l} \ll \dfrac{j\omega L R_p}{R_p + j\omega L}$, we finally get

$$V_m \approx \frac{V_{in}}{10} \left\{ 1 + \frac{\delta r_m}{R_p} + \frac{\delta l_m}{L} + j\left(\frac{\omega\delta l_m}{R_p} - \frac{\delta r_m}{\omega L} \right) \right\} \quad (3.33)$$

Thus, the output voltage at a setting D is

$$V_{out} = V_{in} \left\{ D + \frac{1}{10}\sum_{m=0\cdot1}^{m=D} \left(\frac{\delta r_m}{R_p} + \frac{\delta l_m}{L} \right) + \frac{j}{10}\sum_{m=0\cdot1}^{m=D} \left(\frac{\omega\delta l_m}{R_p} - \frac{\delta r_m}{\omega L} \right) \right\} \quad (3.34)$$

From the definition given in eqn. 3.28, it now follows that the low frequency error is

$$\zeta_{lf} = \frac{1}{10} \sum_{m=0\cdot1}^{m=D} \left(\frac{\delta r_m}{R_p} + \frac{\delta l_m}{L} \right) + \frac{j}{10} \sum_{m=0\cdot1}^{m=D} \left(\frac{\omega \delta l_m}{R_p} - \frac{\delta r_m}{\omega L} \right) \quad (3.35)$$

Remembering the relationships given in eqns. 3.29 and 3.30, it follows that the maximum error occurs at $D = 0\cdot5$ when the 5 lower sections have maximum positive errors and the 5 upper sections have maximum negative errors, or vice versa. If we let δr_{max} and δl_{max} represent the maximum values of δr_m and δl_m, the maximum in phase error is seen to be

$$\zeta_{I\,max} = \frac{\delta r_{max}}{2R_p} + \frac{\delta l_{max}}{2L} \quad (3.36)$$

and, with the worst possible sign combination, the maximum quadrature error is

$$\zeta_{Q\,max} = \frac{\omega \delta l_{max}}{2R_p} + \frac{\delta r_{max}}{2\omega L} \quad (3.37)$$

Variations in the wire diameter cause δr_{max} to be about 1% of $\bar{r}$ and coupling variations cause δl_{max} to be about 2% of $\bar{l}$.[23] Using the typical values given in Section 3.2.3, we find that

$$\zeta_{I\,max} = 7\cdot8 \times 10^{-8} \quad \text{and} \quad \zeta_{Q\,max} = 8\cdot3 \times 10^{-8}$$

When the quadrature error voltage is combined vectorially with the output voltage, its effect on the amplitude of the resultant vector is found to be completely negligible compared with the effect of the in-phase error voltage.

Error caused by core inhomogeneities
Inhomogeneities in the magnetic core cause small deviations in the values of L and R_p, but they are very difficult to assess so they have not been included in the theory given above. However, Hill and Deacon[23] have concluded that the in-phase error due to these deviations is less than 1 part in 10^8.

Error caused by distributed admittances between the windings
At frequencies above 1 kHz, the interwinding capacitances and conductances can cause serious errors. The high-frequency equivalent circuit for a small part of an IVD is shown in Fig. 3.11a. For one stage of a decade divider, the complete equivalent circuit contains 55 capacitances and 55 conductances. The error caused by these distributed admittances can be found by using Kirchhoff's laws and then solving

55 (G + jωC)
99 (G + jωC)
132 (G + jωC)
154 (G + jωC)
165 (G + jωC)
165 (G + jωC)
154 (G + jωC)
132 (G + jωC)
99 (G + jωC)
55 (G + jωC)

Fig. 3.11B Alternative high-frequency equivalent circuit for one decade of an IVD

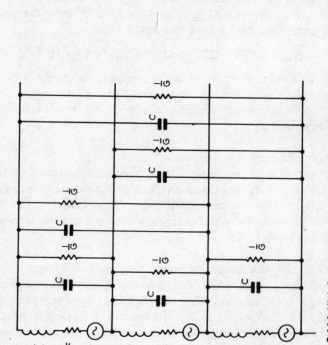

Fig. 3.11A High-frequency equivalent circuit of part of an IVD

t'ie :quations by determinants. The precise details of this process have
l een given by Zapf[19] and Hill and Deacon,[23] and the in-phase error due
·c these distributed admittances is found to be

$$\zeta_D = 91{\cdot}667(2D^3 - 3D^2 + D)(\bar{r}G - \omega^2 C\bar{l}) \tag{3.38}$$

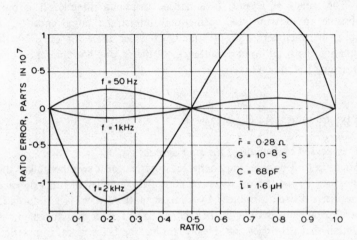

Fig. 3.12 Error caused by distributed admittances between windings

Fig. 3.12 shows how ζ_D varies with D at different frequencies with
typical values for $\bar{r}$, G, C and $\bar{l}$. It should be noted that the error is zero
at any value of D when

$$f = \frac{1}{2\pi} \sqrt{\frac{\bar{r}G}{C\bar{l}}} \tag{3.39}$$

and it is always zero when $D = 0{\cdot}5$. At 1 kHz, with the values given in
Fig. 3.12, the maximum value of ζ_D is seen to be $1{\cdot}3 \times 10^{-8}$. When
$\omega^2 C\bar{l} \gg \bar{r}G$, this error becomes proportional to the square of the
frequency.

As this error is always zero at the central tap, good accuracy can be
obtained up to a high frequency with a binary IVD. Such a device has
been described by Hoer and Smith[22] and its errors are less than 1 part
in 10^7 from 1 kHz to 1 MHz.

Using the results of the analysis referred to earlier, the equivalent
circuit of one decade of an IVD can be redrawn as shown in Fig. 3.11b.
Deacon[27] has pointed out that it is possible to eliminate the distributed
admittance errors by equalising the loading on all sections. To do this,
it follows from Fig. 3.11b that it is necessary to add the following
components:

110 times C and G across sections 1 & 10
66 times C and G across sections 2 & 9
33 times C and G across sections 3 & 8
11 times C and G across sections 4 & 7

Each section is then loaded by $165C$ and $165G$.

The values of C and G cannot be measured directly. It is only possible to measure the open-circuit interstrand admittance $G_m + j\omega C_m$ between any pair of strands of which there are only 45. By equating the total input admittances of the 55 and 45 representations, Deacon[27] has shown that

$$G + j\omega C = 0.68\,(G_m + j\omega C_m)$$

A typical value[27] for C_m is 25 pF/m.

Error caused by loading of the later decades

When up to 3 decades are connected together with switches, all of the coils can be wound on the same core if the number of turns on each successive decade is reduced by a factor of 10. The voltage induced in each paralleling decade is then equal to the voltage induced in the section that it bridges and, ideally, no loading takes place. In practice, various imperfections produce errors greater than 1 part in 10^6, so this technique cannot be used in the early stages of a highly accurate IVD.

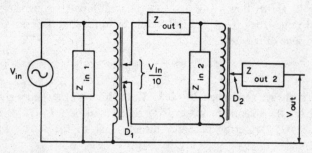

Fig. 3.13 Diagram used for analysis of the error caused by loading of the next decade

When separate cores are used, errors are caused by the finite output impedances and non-infinite input impedances of the autotransformers. To analyse this effect, let us consider the 2-stage IVD shown in Fig. 3.13. In this diagram, the two autotransformers are assumed to be perfect and their imperfections are accounted for by the inclusion of the various Zs. With zero source impedance and infinite load impedance, no errors are caused by Z_{in1} and Z_{out2}, and it is seen that

$$V_{out} = V_{in}D_1 + \frac{V_{in}}{10} \frac{Z_{in2}}{Z_{out1} + Z_{in2}} D_2 \qquad (3.40)$$

With perfect autotransformers, $V_{out} = V_{in}(D_1 + D_2/10)$ so the error due to this cause is seen to be

$$\zeta_L \approx -\frac{D_2}{10} \left\{ 1 - \frac{Z_{in2}}{Z_{out1} + Z_{in2}} \right\} \approx -\frac{D_2}{10} \frac{Z_{out1}}{Z_{in2}} \qquad (3.41)$$

When $D_2 = 1$ and the values given earlier are used, it is seen that ζ_L exceeds 1 part in 10^7. This is the largest source of error yet encountered. However, it can be reduced considerably by using larger cores, more turns, thinner supermalloy tape and two wires connected in parallel throughout the first autotransformer. Another way of reducing this source of error is to use two-stage transformers (see Section 3.2.6).

By simple extension of the theory given in this section, it is easily shown that loading effects in the second interstage are 10 times less important than those in the first interstage, those in the third interstage are 100 times less important and so on. This also applies to division errors in the autotransformers.

Errors caused by impedances of the connecting leads and switches
The errors caused by lead and switch resistances in a Kelvin-Varley voltage divider have already been discussed. Similar errors arise in an IVD, but they can be reduced to a low level by good design and 4 terminal operation. Some commercially available IVDs give an output which is a few parts in 10^7 of the input when all dials are set at zero.

Errors caused by variations in the input voltage, frequency and ambient temperature
Errors due to these causes have been investigated by Hill and Miller[17] on a 7-decade IVD made at the National Physical Laboratory and their findings are summarised below:

Change	Maximum error
Input changed from 5 to 50 V	$< 5 \times 10^{-8}$
Frequency changed from 0·9 to 1·1 kHz	$< 10^{-7}$
Frequency changed from 50 Hz to 2 kHz	$< 10^{-6}$
Temperature changed from 19 to 24°C	$< 2 \times 10^{-8}$

3.2.5 Error when two multi-decade inductive voltage dividers are connected in tandem

When one perfect IVD is set to give a ratio D, the attenuation through it in decibels is given by

$$A = 20 \log_{10} \frac{V_{in}}{D V_{in}} = 20 \log_{10} \frac{1}{D} \qquad (3.42)$$

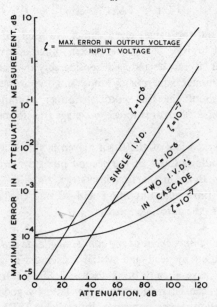

Fig. 3.14 Error due to inductive voltage divider imperfections

If, instead of being perfect, it has a maximum error of ζ, which can be positive or negative, it follows from eqn. 3.28 that

$$V_{out} = V_{in}(D + \zeta) \qquad (3.43)$$

and the attenuation through it in decibels is now given by

$$A' = 20 \log_{10} \frac{V_{in}}{V_{in}(D + \zeta)} \qquad (3.44)$$

Thus, the attenuation error in dB due to ζ is

$$\Delta A' = A' - A = 20 \log_{10} \frac{D}{D + \zeta} \approx 8 \cdot 686 \log_e \left(1 - \frac{\zeta}{D}\right)$$

$$\approx -8 \cdot 686 \, \zeta \, \text{antilog}_{10} \left(\frac{A}{20}\right) \qquad (3.45)$$

Fig. 3.14 shows how $\Delta A'$ varies with A for various values of ζ.

Suppose now that two identical multi-decade IVDs are connected in tandem, as shown in Fig. 3.15, and are set to give ratios of D_3 and D_4 such that $D_3 D_4 = D$. We now have

$$V_{out} = V_{in}(D_3 + \zeta) \cdot \left| \frac{Z_{in}}{Z_{out} + Z_{in}} \right| \cdot (D_4 + \zeta) \qquad (3.46)$$

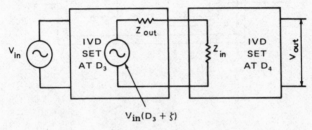

$V_{in}(D_3 + \zeta)$

Fig. 3.15 Diagram used for analysis of the error caused by connecting two IVDs in tandem

Going through a similar process to that given above, the attenuation error in dB is now found to be

$$\Delta A'' \approx 8 \cdot 686 \log_e \left\{ 1 - \frac{\zeta}{D_3} - \frac{\zeta D_3}{D} + \frac{\zeta^2}{D} \right\} + 20 \log_{10} \left| 1 + \frac{Z_{out}}{Z_{in}} \right|$$
$$(3.47)$$

On differentiating $\Delta A''$ with respect to D_3, (ignoring the dependence of Z_{out} on D_3) and then equating to zero, we find that the error is least when $D_3 = D_4 = \sqrt{D}$. Thus,

$$\Delta A''_{min} = -17 \cdot 372 \, \zeta \, \text{antilog}_{10} \left(\frac{A}{40} \right) + 20 \log_{10} \left| 1 + \frac{Z_{out}}{Z_{in}} \right|$$
$$(3.48)$$

The last term represents the loading error that occurs when the two IVDs are connected in tandem. To minimise this error, it is clearly necessary to use IVDs with the lowest possible output impedance and the highest possible input impedance. This error can then be made less than $0 \cdot 0001$ dB. Fig. 3.14 also shows how $\Delta A''_{min}$ varies with A for various values of ζ (negative values of ζ have been used to give the worst case error, and a value of $0 \cdot 0001$ dB has been assumed in each case for the last term). For high values of attenuation, it is clearly advantageous to use two IVDs in tandem. When $\zeta = \pm 10^{-7}$ it can be seen that the cascade arrangement gives an error of less than $0 \cdot 001$ dB at 100 dB.

3.2.6 Two stage inductive voltage dividers

To improve the low-frequency performance of IVDs, considerable attention has been given in recent years to two-stage transformers.[24, 27, 28]

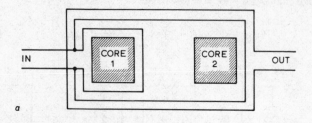

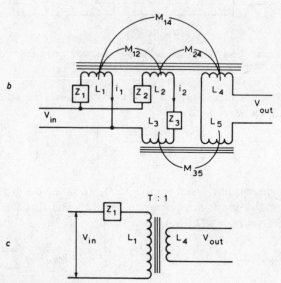

Fig. 3.16 Transformer diagrams

    a Schematic of two-stage transformer
    b Equivalent circuit of two-stage transformer
    c Circuit of corresponding single-stage transformer

A transformer of this type is shown in Fig. 3.16*a* and its equivalent circuit is shown in Fig. 3.16*b*. Two separate magnetic cores and two primary windings are used. One of the primary windings is wrapped around core 1 only, while the other one envelops both cores. The secondary winding also envelops both cores. A full treatment of the two-stage transformer is beyond the scope of this book. However, a simple treatment of it will be given to show its advantages. In Fig.

3.16b, z_1, z_2 and z_3 represent the leakage impedances. Using Kirchhoff's law, we get

$$V_{in} = i_1(z_1 + j\omega L_1) + i_2(j\omega M_{12}) \tag{3.49}$$

$$= i_1(j\omega M_{12}) + i_2(z_2 + z_3 + j\omega L_2 + j\omega L_3) \tag{3.50}$$

$$V_{out} = i_1(j\omega M_{14}) + i_2(j\omega M_{24} + j\omega M_{35}) \tag{3.51}$$

Assuming 100% coupling, we have

$$M_{12}^2 = L_1 L_2, M_{14}^2 = L_1 L_4, M_{24}^2 = L_2 L_4 \quad \text{and} \quad M_{35}^2 = L_3 L_5.$$

To obtain a step down ratio of T to 1, the windings are designed so that

$$L_1 = L_2 \quad \text{and} \quad \frac{L_2}{L_4} = \frac{L_3}{L_5} = T^2$$

To simplify the equations, let $j\omega L_1 = Z_1$ and $j\omega L_3 = Z_3$. Then, from eqns. 3.49 to 3.51, we find that

$$\frac{i_2}{i_1} = \frac{z_1}{z_2 + z_3 + Z_3} \tag{3.52}$$

and

$$\frac{V_{out}}{V_{in}} = \frac{1}{T} \left\{ 1 - \frac{z_1(z_2 + z_3)}{(Z_1 + z_1)(Z_3 + z_2 + z_3) + Z_1 z_1} \right\}$$

$$\approx \frac{1}{T} \left\{ 1 - \left(\frac{z_1}{Z_1 + z_1}\right) \left(\frac{z_2 + z_3}{Z_3}\right) \right\} \tag{3.53}$$

An equivalent one-stage transformer is shown in Fig. 3.16c and in this case, it is immediately seen that

$$\frac{V_{out}}{V_{in}} = \frac{1}{T} \left\{ 1 - \frac{z_1}{Z_1 + z_1} \right\}$$

As z_2 and z_3 are much smaller than Z_3, it is seen that V_{out}/V_{in} is much closer to $1/T$ with a 2-stage transformer than it is with a single-stage transformer.

Fig. 3.17 shows a two-decade IVD that contains 2-stage transformers. The most accurate IVDs that have been described in the literature are based on this principle.

3.2.7 Conclusions

When every source of error in an IVD is reduced to an absolute minimum, an accuracy of a few parts in 10^9 can be achieved. Thus, a top grade IVD is undoubtedly the best variable attenuation reference standard available in the world today. IVDs can be manufactured fairly easily, they are not temperature dependent, they do not require regular recalibration and the only maintenance they need is cleaning of the switch contacts with alcohol about twice a year.

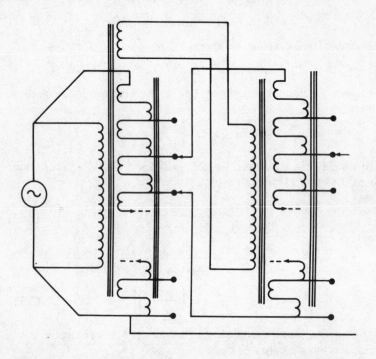

Fig. 3.17 Two-decade two-stage IVD

3.3 I.F. piston attenuators

3.3.1 Introduction
It is recorded that piston attenuators were in use in standard-signal generators for testing radio receivers as early as 1929. Several types were developed and used between then and 1934 when the first paper on the subject was read, and published later in 1935.[38] In the discussion on this paper, the term 'piston attenuator' appeared to be used for the first time. The theoretical work behind the designs was attributed to H.A. Wheeler, but unfortunately was not published; clearly it anticipated by several years the theory of evanescent modes in waveguides (apart from the early paper by Lord Rayleigh). In the 1935 paper, Harnett and Case described three types: variable capacitance (now

termed E_{01} mode in circular guide),* variable mutual inductance with coaxial coils (now termed H_{01} mode in circular guide) and variable mutual inductance with transverse coils (now termed H_{11} mode in circular guide or H_{01} mode in rectangular guide). It is clear that the importance of 'purity of mode' was appreciated in order to obtain a linear decibel scale, and the effect of current penetration on the attenuation rate with mutual inductance coupling was also appreciated. This was very significant at the low frequencies then used (100 kHz to 24 MHz).

Later experience and work in leading standards laboratories on the development of piston attenuators as standards of attenuation working at a fixed frequency between 20 and 60 MHz led to the adoption of the evanescent H_{11} mode in waveguides of circular cross-section. The reasons for this still apply: H_{11} mode has the lowest attenuation rate of all modes so that other modes, if present, will be attenuated more rapidly with distance from the source. Also, a circular section can be produced with greater precision than a rectangular form with its sharp corners, and because precision is of the utmost importance in realizing an attenuation standard, a cylinder of circular cross-section is always used.

The main innovation made almost as soon as the H_{11} mode was adopted was the use of a grid placed across the cylinder in front of the excitation coil or conductors to act as an electric-field screen, to attenuate any E_{01} mode that may be excited.[39,40] This grid is now referred to as a mode filter, has been modified in design and is regarded as something more than an electric field screen.

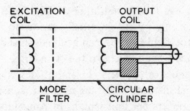

Fig. 3.18 General form of H_{11}-mode piston attenuator

The essential parts of the commonly used form of piston attenuator are shown in Fig. 3.18. The attenuation is varied by altering the distance between the coils, usually by mounting the output coil on a non-contacting piston and threading the output coaxial line through a

---

* Notation in the UK and USA has always differed: *E*-modes are termed *TM* modes and *H*-modes are termed *TE* modes in the USA

hollow piston-rod. Provided the coils and the mode filter are so designed that only one evanescent waveguide mode is present in the cylinder, the voltage induced in the output coil falls exponentially with distance between the coils, at a rate that can be calculated from the measured diameter of the uniform cylinder. When the highest precision is required, a correction depending on the conductivity of the cylinder has to be applied.

Apart from the need to construct a highly conducting cylinder of uniform and accurately known inside diameter that is straight, and to know its high frequency electrical conductivity, the design mostly involves the input and output coils or conductors and their circuits and the mode filter. The need to measure the displacement of the piston with adequate precision does not present a difficult problem nowadays.

3.3.2 Theory of the piston attenuator

The possible distributions of oscillating electromagnetic field inside a conducting cylinder, excited at one end, are derived and stated in Appendix 3. This is an extension of the well-known theory of waveguides to the extreme case where the frequency used corresponds to a free-space wavelength that is very long compared with the diameter $2a$ of the inside of the cylinder. In fact, there is no propagation and the field is termed *evanescent,* which means rapidly diminishing with distance from the source. The field amplitudes all diminish exponentially with distance from the excited end of the cylinder and the E and H components are in time quadrature, so that there is no power transfer inside the cylinder (in the absence of an output coil). It is really no more than a fringe field penetrating weakly down the inside of the cylinder, sustained only by the externally produced field at the input: its virtue is that the field strength can be made an accurately calculable function of distance along the axis of the cylinder.

The axial dependence of the field amplitudes is expressed $e^{-\alpha_{nm}z}$ where z is the axial distance along the cylinder, and α_{nm} is the attenuation constant in nepers per unit length. In Appendix 3 it is expressed as

$$\alpha_{nm} = k \sqrt{\left[1 - \left(\frac{2\pi}{k\lambda}\right)^2\right]} \tag{3.54}$$

where λ is the unbounded wavelength in the same medium as that enclosed by the cylinder, corresponding to the frequency used. k is a constant given by

$$k = \rho_{nm}/a \tag{3.55}$$

where ρ_{nm} is the mth zero of the nth-order Bessel function $J_n(x)$ for E-modes (where only the electric field has an axial component) or the mth zero of the derivative of the nth order Bessel function $J_n'(x)$ for H-modes (where only the magnetic field has an axial component). Some values of ρ_{nm} are:

for E modes: $\rho_{01} = 2 \cdot 40483$, $\quad \rho_{11} = 3 \cdot 83171$, $\quad \rho_{02} = 5 \cdot 52008$,

$\qquad \rho_{12} = 7 \cdot 01559$

for H modes: $\rho_{01} = 3 \cdot 83171$, $\quad \rho_{11} = 1 \cdot 84118$, $\quad \rho_{21} = 3 \cdot 05424$,

$\qquad \rho_{12} = 5 \cdot 33144$

and higher modes have larger values.

For i.f. piston attenuators with an internal diameter $2a$ of a few centimetres, the second term under the root in eqn. 3.54 is negligible, except in the most precise calculations, and $\alpha_{nm} \simeq k = \rho_{nm}/a$. In any case the attenuation rate is proportional to ρ_{nm}. It is clear that the H_{11} mode has the lowest rate of attenuation and that at a sufficient distance from the source-end this mode will dominate any other modes that might also have been excited. The number of modes excited and their initial relative amplitudes and phases depend on the form of the excitation field. By suitable design of the excitation conductors or coil, the H_{11} mode can be made predominant initially. Use of a mode filter can attenuate any E_{01} and E_{11} modes considerably, and any H_{01} and H_{21} modes partially. Thus, the field in the cylinder can be substantially a pure H_{11} mode with a simple law of attenuation $e^{-\alpha z}$ for the field amplitudes with distance z along the cylinder. It is essential that this is achieved over all the usable range of attenuation.

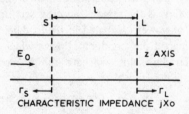

Fig. 3.19 Planes of reference and reflection coefficients in cylinder

Let us consider a pure H_{11} mode in the cylinder and refer to Fig. 3.19. Let S be a cross-section suitably near the exciting coil, loop or rods (the 'source') and let L be another cross-section distant l from S, suitably near the output coil or loop (the 'load'). To provide an amplitude reference, let E_0 be the electric field intensity (V/m) at the centre of the cylinder at S, and let V_0 be the r.m.s. voltage across a certain diameter at S corresponding to E_0. It is shown in Appendix 3 that, for the H_{11} mode,

$$V_0 = 1 \cdot 74 a E_0 \tag{3.56}$$

This voltage, together with the 'imaginary power' (or rather reactive volt-amperes in the cylinder) can be used to define a characteristic impedance

$$Z_0 = jX_0 = j2 \cdot 03 \frac{\omega\mu}{\alpha_{11}} \tag{3.57}$$

and if $\alpha_{11} \simeq k_{11}$ we have

$$jX_0 \simeq j1 \cdot 1 \ \omega\mu a \quad \text{(ohms)} \tag{3.58}$$

where a is the radius of the cylinder. With these definitions, the following expression for the power absorbed in the 'load' at L is obtained in Appendix 3, Section 3.4:

$$P_L = \frac{|V_0|^2 j(\Gamma_L - \Gamma_L^*) e^{-2\alpha l}}{X_0 |1 - \Gamma_S \Gamma_L e^{-2\alpha l}|^2} \tag{3.59}$$

where Γ_S and Γ_L are 'voltage' reflection coefficients (relative to the imaginary characteristic impedance jX_0) of the 'source' and the 'load', respectively. Notice that unless Γ_L has an imaginary component there can be no power transfer from source to load through the cylinder. It has already been stated that a single evanescent field transfers no power and has a purely imaginary characteristic impedance, being a purely reactive or 'storage' oscillating field. Only when a 'reflected' field with a suitable phase lag in the reflection is superimposed is there any power flow. When such superposition of evanescent fields takes place, the power flow is the same at all cross-sections of the cylinder, only the amount of power dissipated in the load being drawn from the source. This is apart from conductor losses which are for the most part negligible. To show this, we have for the power density at an arbitrary point distant z from S:

$$\underbrace{jE_0 H_0 e^{-2\alpha z}}_{\substack{\text{incident reactive 'volt} \\ \text{amperes' per } m^2}} \cdot \underbrace{j2|\Gamma_L| e^{-2\alpha(l-z)} \sin\psi}_{\substack{2 \times \text{imaginary part of reflection coefficient} \\ \text{referred to position } z}}$$

$$= \underbrace{-2E_0 H_0 e^{-2\alpha l} |\Gamma_L| \sin\psi}_{\text{real power (independent of } z) (\text{W/m}^2)}$$

which follows from eqn. 3.59 with $\Gamma_S = 0$ and with $|V_0|^2/X_0$ replaced here by $E_0 H_0$. For power to be absorbed by the load, ψ must be between 0 and $-\pi$ with an optimum at $-\pi/2$ for which $\sin \psi = -1$. This corresponds to a resistive load. A positive value of ψ between 0 and π would require a negative resistance in the load.

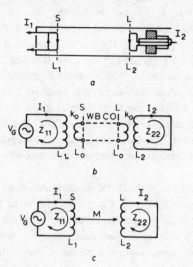

Fig. 3.20 Piston attenuator
   a Schematic
   b Equivalent circuit
   c Incomplete equivalent circuit

The next important question to be answered is how to calculate the coupling between the magnetic fields of the excitation and output coils and the field in the cylinder, so that the voltage V_0 across the cylinder at S can be related to a source and the voltage reflection coefficients Γ_S and Γ_L can be related to the properties of the circuits coupled to the cylinder at S and L, respectively. Let the fraction k_0 of the magnetic flux generated by the current I_1 in the coil of inductance L_1 (Fig. 3.20*a*) form the initial magnetic flux in the cylinder. To line up with the scheme envisaged in eqn. 3.59 and shown in Fig. 3.19, it is necessary to describe this initial magnetic flux in terms of a voltage across a diameter of the cylinder (in a direction at right angles to the axis of the coil) and this is done in terms of the characteristic impedance of the cylinder (for the H_{11} mode) which is that of an inductance $L_0 = 1 \cdot 1 \, \mu a \dots (H)$ by eqn. 3.58.

Thus the equivalent input circuit used is that shown on the left of Fig. 3.20*b* where L_1 is coupled to an inductor L_0 which is connected directly across the equivalent circuit of the cylinder at S. In this way, the lumped-circuit network is joined to the waveguide 'network' at the plane S.

The magnetic flux linking the inductor L_0 is the same as the flux entering the cylinder, which on account of the characteristic impedance $j\omega L_0$ of the cylinder produces the voltage V_0 at S, (provided the cylinder is very long or it is matched at the far end to the right). The impedance seen looking back to the left at S in Fig. 3.20a is that of the continuation of the cylinder to the left of S, modified by any reflection from the end, and by the circuit coupled through L_1. If L_1 is open-circuited, and the far end of the cylinder is distant at least a from S, the effect of reflection from the end is small and the impedance seen to the left of S is $j\omega L_0$, the characteristic impedance of the cylinder. This also is the value obtained from the equivalent circuit in Fig. 3.20b. Therefore, in either direction, the equivalent circuit gives a correct representation. A similar equivalent circuit represents the load coupling at L.

The impedance Z_{11} is the total mesh impedance of circuit 1 when L_0 at S is open-circuited, and Z_{22} is the total mesh impedance of circuit 2 when L_0 at L is open-circuited, as is standard procedure for lumped circuits involving mutual couplings. V_G is the r.m.s. voltage of a generator of zero internal impedance, any actual generator impedance being included in Z_{11}.

Earlier work adopted the arrangement shown in Fig. 3.20c to represent the piston attenuator,[41,42] but although such an arrangement gave enough information to help estimate the interaction between source and load at close coupling, the resulting theory could not be lined-up with the waveguide theory leading to eqn. 3.59. The presence of the metal cylinder constitutes a link circuit, even for the lowest frequencies, and Fig. 3.20c is not a complete representation of the piston attenuator.

Reffering to Fig. 3.20b, the effective load impedance at L is found to be

$$Z_L = j\omega L_0 + \frac{\omega^2 k_0^2 L_2 L_0}{Z_{22}} \tag{3.60}$$

and the effective source impedance at S is

$$Z_S = j\omega L_0 + \frac{\omega^2 k_0^2 L_1 L_0}{Z_{11}} \tag{3.61}$$

The Thévenin-equivalent voltage generator at S is $V_S = j\omega k_0 \sqrt{L_1 L_0} \cdot I_1$ or

$$V_S = \frac{j\omega k_0 \sqrt{L_1 L_0}}{Z_{11}} V_G \tag{3.62}$$

The relation between V_S and V_0 is readily found to be

$$V_0 = \frac{V_S j X_0}{Z_S + j X_0} = \tfrac{1}{2} V_S (1 - \Gamma_S) \tag{3.63}$$

where

$$\Gamma_S = \frac{Z_S - j X_0}{Z_S + j X_0} \tag{3.64}$$

and X_0 is written for ωL_0. Likewise,

$$\Gamma_L = \frac{Z_L - j X_0}{Z_L + j X_0} \tag{3.65}$$

Use of eqns. 3.62 and 3.63 in eqn. 3.59 leads to the relation

$$P_L = \frac{|V_G|^2}{|Z_{11}|^2} \cdot \frac{\omega k_0^2 L_1}{4} \cdot \frac{|1 - \Gamma_S|^2 j (\Gamma_L - \Gamma_L^*) e^{-2\alpha l}}{|1 - \Gamma_S \Gamma_L e^{-2\alpha l}|^2} \tag{3.66}$$

for the power absorbed in the load, where l is the distance between the faces of the excitation and output coils. Γ_S and Γ_L are evaluated from eqns. 3.64 and 3.65, using eqns. 3.61 and 3.60 for Z_S and Z_L. Again, Fig. 3.20b refers.

This equation is an expression of the theory of the piston attenuator. From it, one can derive the relative attenuation associated with two settings l_1 and l_2 of the piston. Thus:

$$\frac{P_{L1}}{P_{L2}} = e^{2\alpha(l_2 - l_1)} \left| \frac{1 - \Gamma_S \Gamma_L e^{-2\alpha l_2}}{1 - \Gamma_S \Gamma_L e^{-2\alpha l_1}} \right|^2 \tag{3.67}$$

or in decibels, $10 \log_{10}(P_{L1}/P_{L2})$ where $P_{L1} > P_{L2}$ and $l_2 > l_1$, we have

$$A = 8 \cdot 686 \alpha (l_2 - l_1) + 10 \log_{10} \left| \frac{1 - \Gamma_S \Gamma_L e^{-2\alpha l_2}}{1 - \Gamma_S \Gamma_L e^{-2\alpha l_1}} \right|^2 \quad \text{(dB)} \tag{3.68}$$

The first term shows an attenuation linearly dependent on the displacement between the two coils, while the second term results from *interaction between the source and the load*. This second term approaches zero when both l_1 and l_2 are large, and it can be made less than any assigned amount, even for l_1 or l_2 small, by making the product $\Gamma_S \Gamma_L$ small enough. To do this, Γ_S must effectively be reduced to zero because, as we have seen already, Γ_L must have a finite (imaginary) component if any power at all is to be transferred from source to load. So special means such as stabilisation of the current I_1 in the coil L_1

must be used. This is so because the piston attenuator attenuates wholly by reflection, so that the load impedance seen from S is bound to vary as the attenuation is varied, and $\Gamma_S \Gamma_L$ cannot be made zero by matching in the conventional sense as with real characteristic impedance in a transmission guide. Stabilisation of I_1 to a fixed value is equivalent to making Z_{11} infinite, (and V_G infinite also to make $V_G/Z_{11} = I_1$ finite) in Fig. 3.20b. Then from eqn. 3.61, $Z_S = j\omega L_0 = jX_0$ and from eqn. 3.64, $\Gamma_S = 0$. In so far as this can be achieved, the second term in eqn. 3.68 vanishes.

Other means, such as tuning adjustments to the circuits 1 and 2 have been used to minimise the non-linearity resulting from the interaction.[41] When the i.f. source is not integral with the piston attenuator, stabilisation of I_1 is not normally possible (indeed, the piston attenuator may then be used with a variable source) and such means for minimising interaction are very useful.

Eqns. 3.66, if used, 3.67 and 3.68 are best left in their present forms, calculating the complex Γ_S and Γ_L by eqns. 3.64 and 3.65 in specific examples, rather than expressing the relations directly in terms of the components of Z_{11} and Z_{22}, L_1, L_2 and so on.

So far, nothing has been said about the value of the coupling coefficient k_0. To a first approximation, if the coil has a length of cylinder on both sides, $k_0 \simeq \frac{1}{2}$ because the coil couples equally to the evanescent fields on either side of it, while the cylinder contains all the field of the coil. If a fraction of the magnetic field from the coil couples to modes other than H_{11} in the cylinder, then, for the H_{11} mode, k_0 would be less than $\frac{1}{2}$ by a small amount.

3.3.3 Excitation and purity of mode

Evanescent H-modes are excited by current carrying conductors. The magnetic field in the immediate neighbourhood of the conductors can be matched more or less by a mixture of the magnetic fields of few or many evanescent H-modes in various proportions with given relative phases. Likewise, any electric field near the conductors may be matched with evanescent E-modes, if they fit. But quite often the electric field of the conductors agrees with the electric field associated with the magnetic field of the dominant evanescent H-mode stimulated by the current in the conductors. As already mentioned, the aim is to obtain a pure H_{11} mode in the cylinder, and the discussion will be limited to the requirements of this aim.

The ideal way to produce a pure H_{11} mode is by a current sheet[42] with the distributions of strength and direction of the current like those of the electric field intensity of the mode. This can be shown as follows.

Let I be the current per unit width of sheet, with components I_r and I_ϕ in cylindrical coordinates. Then in Fig. 3.21a, the current for the width $rd\phi$ is $I_r rd\phi$ and this is equal to the integral of the resulting magnetic field around the current strip, i.e. from the figure $-2H_\phi rd\phi$. Again, in Fig. 3.21b, the current for the width dr is $I_\phi dr$ and this equals the integral of the resulting magnetic field around the strip, i.e. $2H_r dr$. So

$$I_r = -2H_\phi \quad \text{and} \quad I_\phi = 2H_r \tag{3.69}$$

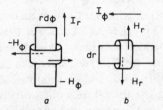

Fig. 3.21 Components I_r and I_ϕ of current sheet, with resulting magnetic field intensities and directions
a Current for width $rd\phi$
b Current for width dr

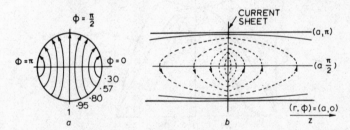

Fig. 3.22 Electric and magnetic fields for H_{11} mode excited by current sheet
a Transverse section showing electric field
b Longitudinal section showing magnetic field. Directions and relative densities of current as shown in *a*

Thus $I_r/I_\phi = -H_\phi/H_r$. But, as shown in Appendix 3, Section A3.2, $-H_\phi/H_r = E_r/E_\phi$ so that $I_r/I_\phi = E_r/E_\phi$. It follows from this last relation and the argument in A3.2 that the current flow lines and the lines of electric force are the same. These are shown in Fig. 3.22a which is based on Fig. A3.1 of Appendix 3. From the second part of eqn. 3.69 and the expression for H_r given in eqn. 46 of Appendix A3.3, the intensity of current across the radius $\phi = 0$ varies as $J_1'(kr) = J_1'(1\cdot841 \ r/a)$. It is a maximum at the centre and zero at the cylinder wall, relative current density values being shown on the lines in Fig. 3.22a applicable strictly where they cross the diameter at $\phi = 0$. It should be noted that as the current lines approach the cylinder wall, the width associated with a given current strip increases so that the current in the strip is constant along its length. If a current sheet of this form could be realised the magnetic field in the plane $\phi = 0$ along the z-axis would be like that shown in Fig. 3.22b, with similar evanescent fields to right and left of the current sheet (but with mirror dependence on variables). The mode would be pure H_{11}. It is necessary to continue the cylinder behind the exciting plane to accommodate the field at the back and produce substantially a matched source impedance jX_0.

A serious attempt to achieve an approximation to such a current sheet was recorded by Allred and Cook[43] in 1960, who indeed first drew attention to the ideal way to produce a pure H_{11} mode. This first attempt did not satisfy the $J_1'(kr)$ dependence of current density across the diameter through $\phi = 0$, and an improved design with a thick straight conductor down the centre with two curved thin conductors on eigher side was found to be a much better approximation, described in 1971.[42]

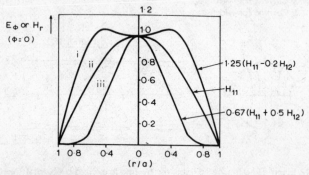

Fig. 3.23 Field strength distribution
  i Across cylinder for coil of length 0·8*a*
  ii Across ideal current sheet
  iii Across single rod perpendicularly across cylinder
   Field strength represented approximately by combinations of H_{11} and H_{12} modes

In general, one or more conductors that are more or less parallel across the cylinder, but not touching the cylinder wall, produce a field distribution that is a combination of the H_{11} and H_{12} modes with some H_{1m} modes of higher order. A good approximation and a workable theory result from consideration of the H_{11} and H_{12} modes alone. Thus, in Fig. 3.23, the broad curve represents $1.25 (H_{11} - 0.2 H_{12})$ which approximates to the field strength distribution of a coil on a rectangular former with axis of length about 0·8*a* across the cylinder. The narrow curve represents $0.67 (H_{11} + 0.5 H_{12})$ which approximates to the field strength distribution of a single straight rod along the diameter through $\phi = \pi/2$. The medium curve is for H_{11} alone. These approximations give an important clue to relative magnitudes and phases of the initial H_{12} mode under typical excitation conditions. Although this H_{12} mode has a high attenuation rate ($\rho_{12} = 5.331$), it is significant because it is not appreciably attenuated by the mode filter used to attenuate the otherwise troublesome E_{01} mode.

The E_{01} mode, which has an attenuation rate only a little higher than that of the H_{11} mode ($\rho_{01} = 2.405$), has received more attention in the past than any other unwanted mode. Detailed methods for detecting and measuring the parameters associated with any E_{01} mode present, using a rotatable pick-up loop, were worked out by R.N. Griesheimer[44] based on earlier work by S.G. Sydoriak in 1943. The importance of the E_{01} mode is offset by two things. First, at frequencies up to 60 MHz the excitation and output coils or loops have low impedances while the E-modes have high impedances, which results in weak coupling. However, at higher frequencies, especially at microwaves, the coupling is stronger. Second, mode filters can be constructed which attenuate the E_{01} (and other E modes) to the point of near extinction while not attenuating the H_{11} mode appreciably. The mode filter usually employed nowadays is a metal grating placed across the cylinder and connected to it, a short distance ($\frac{1}{2}a$ to a is typical) in front of the excitation conductors. Experimental results[42] seem to show that the maximum attenuation of the E_{01} mode occurs when the depth of strip of an edge-on strip grating is about 1.64 times the pitch of the grating, for a frequency of 30 MHz. In such a case the attenuation of an E_{01} mode can be over 80 dB while that of the H_{11} mode can be as low as 0.2 dB. Wire gratings have less effect.[40] The orientation of the mode filter is shown in Fig. 3.25.

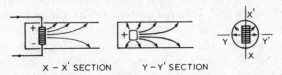

X – X' SECTION   Y – Y' SECTION

Fig. 3.24 E_{01} mode produced by unbalanced coil

With an unbalanced coil or rod for excitation, the E_{01} mode lines of electric force look initially like those shown in Fig. 3.24. An unbalanced output coil or loop can respond to this mode, but again the high impedance of the mode and the low impedance of the coil or loop results in weak coupling to the E_{01} mode.

With a balanced coil or rod, an E_{11} mode may be excited, which initially looks like that in Fig. 3.26. A balanced or unbalanced pick-up coil or loop can respond to this mode, according to circumstances, but the coupling is likely to be weak.

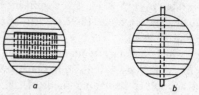

Fig. 3.25 Orientation of mode-filter and coil or rod, to produce minimum H_{11} attenuation and maximum E_{01} attenuation
a Coil excitation
b Rod excitation

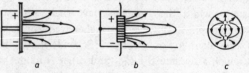

Fig. 3.26 E_{11} mode produced by balanced coil or rod
a Rod
b Coil
c Section

By unbalance between the two halves of a rod or coil with grounded centre-tap, small amounts of H_{01} and H_{21} modes can be produced. But the coupling of the output coil or loop to these modes is likely to be small and these modes are unlikely to be significant in a sensibly designed piston attenuator.

Suppose now that a mode filter is used to reduce any E_{01} or E_{11} modes to a negligible amount and that the excitation system is reasonably symmetrical and balanced so that modes like H_{01} and H_{21} are negligible. Then all that remains are the wanted H_{11} mode and some H_{12} mode, perhaps with smaller amounts of H_{1m} modes ($m > 2$). Therefore, as two examples of the combination of two modes at various distances from the excitation conductors, we consider the H_{11} and H_{12} modes in the proportions and relative phases illustrated in Fig. 3.23. It should be noted from eqns. 44 to 47 of Appendix 3 that the ratio of transverse E to H, $E_r/H_\phi = -E_\phi/H_r = j\omega\mu/\alpha$ (field or wave impedance) is the same for H_{11} and H_{12} modes. Then, if the output coil or loop embraces only the centre part of the magnetic field of the H_{12}

mode (see Fig. 3.27), the coupling of the coil or loop is the same for the H_{11} and the H_{12} mode fields and the power received is proportional to $|E|^2$ because the impedance is the same for both modes.

If two amplitudes E_1 and E_2 which differ in phase by θ are combined, it is well known that the resultant amplitude E is given by

$$E^2 = E_1^2 + E_2^2 + 2E_1E_2 \cos \theta$$

If we write $E_1 = e^{-\alpha_1 z}$ and $E_2 = qe^{-\alpha_2 z}$ for the H_{11} and H_{12} modes, respectively, at z, then:

$$
\begin{aligned}
E^2 &= e^{-2\alpha_1 z} + q^2 e^{-2\alpha_2 z} + 2q e^{-(\alpha_1 + \alpha_2)z} \cos \theta \\
&= e^{-2\alpha_1 z} [1 + q^2 e^{-2(\alpha_2 - \alpha_1)z} + 2q e^{-(\alpha_2 - \alpha_1)z} \cos \theta] \\
&= e^{-2\alpha_1 z} [1 + 2q e^{-(\alpha_2 - \alpha_1)z} \cos \theta]
\end{aligned}
$$

as $z \geqslant a$ in a practical i.f. piston attenuator. Now power is proportional to E^2, so, in decibels, $-A = -4 \cdot 343 (2\alpha_1 z) + 4 \cdot 343 \ln [1 + 2q \cos \theta \, e^{-(\alpha_2 - \alpha_1)z}]$ and as $\ln (1 + x) \approx x$ when $x \ll 1$, we have

$$A = 8 \cdot 686 \, \alpha_1 z - 8 \cdot 686 \, q \cos \theta \, e^{-(\alpha_2 - \alpha_1)z} \text{ (dB)} \qquad (3.70)$$

We have $\alpha_1 = 1 \cdot 841/a$ and $\alpha_2 = 5 \cdot 331/a$ so that $(\alpha_2 - \alpha_1) = 3 \cdot 49/a$. Then for the short coil, $q = 0 \cdot 2$ and $\theta = \pi$, so

$$A(\text{dB}) = 15 \cdot 99 \frac{z}{a} + 1 \cdot 737 \, e^{-3 \cdot 49z/a} \qquad (3.71)$$

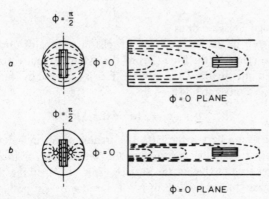

Fig. 3.27 Short pick-up coil in H_{11} and H_{12} mode fields
    a H_{11} mode
    b H_{12} mode

For the single rod exciter and single loop or short coil output, $q = 0 \cdot 5$ and $\theta = 0$, so
$$A(\text{dB}) = 15 \cdot 99 \frac{z}{a} - 4 \cdot 343 \, e^{-3 \cdot 49z/a} \qquad (3.72)$$

The resulting departures from linearity are shown in Fig. 3.28, *a* and *b*, where the 'zero' on the scale is taken as $z = a$. It is clear that the worst case is that of the single rod exciter and single loop output the result being that of Fig. 3.28*b*. This situation can be improved by using a long output coil so that (Fig. 3.27*b*) parts of the two side components of the magnetic field of the H_{12} mode are picked up as well as the centre part. As the centre and side parts are in opposition, the H_{12} mode at the output can be made negligible by choice of length of output coil.

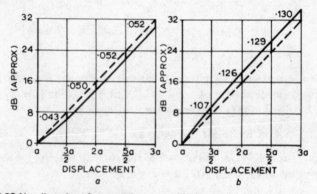

Fig. 3.28 Non-linearity of attenuation versus displacement relations
  a For coil
  b For rod represented by fields of Fig. 3.23

3.3.4 Design considerations

The first step in the design of an i.f. piston attenuator using the H_{11} mode is to calculate the bore ($2a$) of the cylinder required to obtain a given rate of attenuation (α, neper/m). This is achieved by the following relation based on eqn. 3.54:

$$a = \rho_{11}/\sqrt{\{\alpha^2 + (2\pi/\lambda)^2\}} \tag{3.73}$$

but noting that *a* here is the 'effective' radius of the cylinder which is slightly larger than the measured radius on account of the effect of current penetration below the surface. This effect of the finite conductivity was studied by J. Brown[45] and in the present problem, for all practical purposes, the result shows that the measured radius of the cylinder is $(a - \delta/2)$ where δ is the conventional 'skin depth' of the current in the conductor.

$$\delta = 1/\sqrt{(\pi\mu f\sigma)} \tag{3.74}$$

where $\mu = 4\pi \times 10^{-7}$ H/m, f is in Hz and σ is the conductivity in S/m.

If the conductor is of magnetic material, such as stainless steel, then μ is larger. To use eqn. 3.73 it is necessary to convert decibels into nepers very precisely. The factor is 0·11512926. Also, $\rho_{11} = 1\cdot8411838$ and the velocity of propagation in air required to calculate λ from f with sufficient accuracy is $2\cdot9979 \times 10^8$ m/s. The effect of the term in λ is quite small with i.f. piston attenuators.

A difficulty usually arises in the determination of the effective value of σ for use in eqn. 3.74. For metals such as copper or silver, the effective value is generally smaller than the value found from d.c. measurements. Where an accuracy of the order of 0·001 dB per 10 dB is sought, an accurate estimate of δ for the actual cylinder used is needed. One method is to put conducting ends on the cylinder and use it as a resonator in the H_{111} mode. Then $\delta/2$ can be related to the measured resonant frequency. Of course, this resonant frequency is many times higher than the i.f. at which the piston attenuator is to be used, and the value of δ needed is obtained from the relation $\delta \propto f^{-1/2}$.

The cylinder can be made in one of at least two ways. Either it can be electroformed on a stainless steel former, or it can be specially bored out of a thick brass tube. In either case the bore must be measured in detail, usually by an air-gauge or a capacitance gauge which is calibrated against a standardised ring gauge. The bore must be straight and uniform, and, if there is any slight ellipticity, the direction of the major axis must not twist with distance along the axis of the cylinder. If the required accuracy is $\pm0\cdot001$ dB per 10 dB and the bore radius is 10 mm, this radius must be held to within $\pm1\,\mu$m.

The piston attenuator arrangement can be designed in several possible ways. The most common way is shown in Fig. 3.29a where the excitation coil and mode filter are fixed and the output coil is mounted in a non-contacting piston which is displaced to vary the attenuation. This requires a flexible coaxial cable to connect the pick-up coil to the next i.f. stage. Such a cable must be well screened, preferably with a continuous (but flexible) outer conductor, and when it is flexed, the propagation constants of the cable may change. The result is an uncertainty in attenuation possibly of the order of $\pm0\cdot01$ dB or more.

When such an uncertainty is not acceptable it is possible, by light weight components and compact solid-state devices, to build the i.f. source oscillator and amplifier in the back of a non-contacting piston (Fig. 3.29b) which also houses the exciter coil. Then only d.c. supply leads have to be flexed as the attenuation is varied. Such a source will be of limited power and a multi-turn single layer excitation coil will be needed to obtain sufficient field-strength in the cylinder.

If the piston attenuator is to be used as a variable insertion-loss device, then the i.f. source cannot be built-in and either the arrangement in Fig. 3.29*a* is used, or that in Fig. 3.29*c* which employs a line stretcher has to be used, to keep the length between the input and output coaxial connectors constant. The design of the line stretcher requires careful attention if errors are not to be introduced.[47] In any of these arrangements, the mode filter can be near either of the coils, or a mode filter can be fitted near both coils.[49]

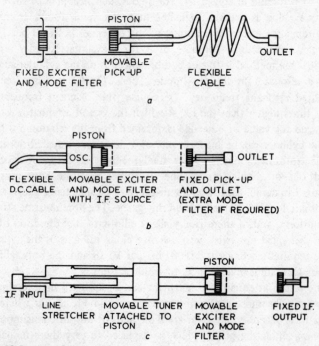

Fig. 3.29 Designs of piston attenuator

    a Conventional
    b With source attached to piston containing excitation coil
    c With fixed output and coaxial line stretcher to keep overall length constant

All pistons should be non-contacting, usually with a thin coating of PTFE (Telfon) to form the bearing. Air bearings are used in the NPL standard piston attenuator.[48]

Measurement of the displacement of the piston is either by magnified viewing of marks on a metal or glass scale and vernier, or by optical methods using a grating, photocell and counter, or using a laser interferometer with an electronic counter. The method used depends on the precision required.

Design of the excitation system and mode filter has been considered in the previous section; now the design of the output system must be considered.

Apart from the interaction term, it is seen from eqn. 3.59 that the power transferred to the 'load' in the output circuit is proportional to $j(\Gamma_L - \Gamma_L^*)$ which must be real and positive. The reflection coefficient here is related to the impedance Z_L by eqn. 3.65 and the requirement that $j(\Gamma_L - \Gamma_L^*)$ be real and positive means that Z_L must have a resistive component. The output circuit is usually like that shown in Fig. 3.30 where L_2 is the inductance of the coupling coil or loop, C_2 is a series capacitor for tuning out all or part of the inductive reactance present with loose coupling, and R_2 is the load resistance.

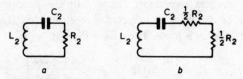

Fig. 3.30 Assumed equivalent circuit of output with pick-up coil, tuning capacitor and load resistor
a Basic
b With fixed resistor associated with pick-up coil to make effective source impedance match load $(R_2/2)$ when L_2 and C_2 are series resonant

To obtain an acceptable value of $j(\Gamma_L - \Gamma_L^*)$, the imaginary part of Γ_L must be negative and not too small. The real part of Γ_L plays no part in this, but the real part of Z_L contributes strongly to the imaginary part of Γ_L. In Reference 46, Fig. 142(*a*) shows a circle diagram for positive imaginary characteristic impedance of the form jX_0 encountered here. Some readers may find it helpful to use such a diagram to illustrate some of the points made here. For the circuit in Fig. 3.30, the load impedance (normalised to X_0) can be derived using eqn. 3.60 and is expressed as

$$z_L = \frac{Z_L}{X_0} = \frac{K_1 + j(1 + K_2^2 - K_1 K_2)}{1 + K_2^2} \quad (3.75)$$

where
$$K_1 = k_0^2 \omega L_2 / R_2, K_2 = X_2 / R_2 \quad \text{and} \quad X_2 = \omega L_2 - 1/\omega C_2$$

With an imaginary characteristic impedance, the modulus of Γ_L is not restricted to less than unity as it is for a real characteristic impedance, but it can have any positive value. Nevertheless, it is seen from eqn. 3.59 that too large a value of Γ_L will increase the interaction by an unacceptable amount. This point is discussed in Reference 46, where it is concluded that a maximum of 1 should be imposed on $|\Gamma_L|$. This arbitrary maximum is obtained if, in eqn. 3.75, $K_1 = 2$ and $K_2 = 1$, so $z_L = 1$ and by eqn. 3.65, $\Gamma_L = -j$ so $j(\Gamma_L - \Gamma_L^*) = 2$. For 30 MHz, the component values for this are with $k_0 = \frac{1}{2}$ and $R_2 = 100\,\Omega$, $L_2 = 4\cdot2\,\mu H$, $C_2 = 7\cdot6\,pF$. To obtain something approaching source match as seen from the receiver, the load resistance is often split as shown in Fig. 3.20b, with a resistor $\frac{1}{2}R_2$ acting as part of the 'source' from the viewpoint of the receiver, which has an input resistance $\frac{1}{2}R_2$. But even with this, the values of L_2 and C_2 obtained above lead to a 'source' impedance $50 + j100\,\Omega$ at the end of a $50\,\Omega$ transmission line. Also, the value $4\cdot2\,\mu H$ is too large to be obtained easily with a single-layer coil on a rectangular former. It is better to have a smaller value of L_2 and to tune C_2 to series resonance. Then, with the arrangement of Fig. 3.30b, a source match is seen from the receiver when the coupling in the piston attenuator is loose. Thus, $K_2 = 0$,

$$z_L = K_1 + j \quad \text{giving} \quad \Gamma_L = \frac{K_1^2 - 2jK_1}{K_1^2 + 4} \quad \text{or}$$

$$j(\Gamma_L - \Gamma_L^*) = \frac{4K_1}{K_1^2 + 4} \tag{3.76}$$

Then, if $\omega L_2 = 2R_2$, with $k_0 = \frac{1}{2}$ (i.e. $L_2 \simeq 1\,\mu H$) we obtain $K_1 = \frac{1}{2}$ and $j(\Gamma_L - \Gamma_L^*) = 0\cdot47$. Smaller values of L_2 give smaller coupling values, roughly in proportion.

Other arrangements of receiver and input circuits can be used, such as one with series tuning the input, and parallel tuning the output,[41] with definite aims in view. The design is not critical, and a wide range of different values of L_1 and L_2 can be made to work with greater or lesser success.

3.3.5 Summary of sources of error

(i) Eqn. 3.54 shows when λ is large that $\alpha \simeq \rho_{11}/a$ and differentiation leads to:

$$\frac{\Delta\alpha}{\alpha} = -\frac{\Delta a}{a} \tag{3.77}$$

It follows that the tolerance on the radius (or diameter) of the cylinder must be about ± 1 part in 10^4 to achieve an error in the rate of $\pm 0\cdot001$ dB per 10 dB.

(ii) To avoid erratic behaviour, a non-contacting piston should be used.

(iii) Non-linearity of attenuation change versus displacement of the piston can be checked by careful measurement by the piston attenuator of a fixed attenuation of, say, about 30 dB at several signal levels, so that the maximum level used corresponds to close spacing and more distant spacings between the coils of the piston attenuator. Non-linearity has two main causes:

(a) the excitation of unwanted evanescent modes, some of which can be reduced to a negligible amount by the use of an effective mode filter

(b) interaction between excitation and output coils which varies with the tightness of coupling. In a piston attenuator with a built-in source, this interaction can be avoided to a great extent by stabilising the current in the excitation coil, loop or rods. However, for a piston attenuator used as a variable insertion loss only very careful design of the input and output circuits and the looseness of coupling can bring the interaction error within suitably chosen limits. This is a direct result of the fact that the piston attenuator attenuates by reflection, and the input and output impedances are functions of the attenuation, marked changes in the input and output impedances occurring when the coils approach one another closely. For this reason, source and load impedances must be constant while the piston attenuator is varied (to produce a variable insertion loss) and should be matched carefully to the nominal characteristic impedance.

(iv) A flexing coaxial cable carrying i.f. can vary in its transmission properties, resulting in an uncertainty of about ±0·01 dB or more. This occurs in designs like Fig. 3.29a. To avoid the uncertainty, the arrangement of Fig. 3.29b can be adopted, or the equivalent arrangement like (a) with the i.f. source attached to the cylinder and having a flexible d.c. cable. The piston is fixed and the i.f. source plus cylinder are moved to vary the attenuation. The flexible i.f. cable is unnecessary here. This latter arrangement was suggested by D. Woods.

When the whole arrangement is used as a variable i.f. insertion loss, a line stretcher is used as in Fig. 3.29c. The line stretcher must be of large diameter and free from erratic contact behaviour, otherwise it would have no advantage over a flexible cable.[47]

3.4 References

1 VARLEY, C.F.: 'On a new method of testing electric resistance', *Math. & Phys. Sect. Brit. Assoc. Adv. Sci.*, 1866, pp. 14–15

2 KELLEY, J.B., and MAROLD, H.H.: 'The Kelvin-Varley voltage divider', *Instrum. & Control Syst.* 1960, **33**, pp. 626–628

3 MORGAN, M.L., and RILEY, J.C.: 'Calibration of a Kelvin-Varley standard divider', *IRE Trans.*, 1960, **I-9**, 237–243

4 FRYER, L.C.: 'A voltage divider standard', *AIEE Trans.*, 1962, **81**, pp. 128–135

5 DUNN, A.F.: 'Calibration of a Kelvin-Varley voltage divider', *IEEE Trans.*, 1964, **IM-13**, pp. 129–139

6 NEFF, S.H.: 'An inexpensive precision voltage divider', *Am. J. Phys.*, 1970, **38**, pp. 769–771

7 BLUMLEIN, A.D.: British Patent 323037, 1928

8 WAGNER, K.W.: 'Zur Messung dielektrischer Verluste mit der Wechselstrom-brucke', *Elektrotech. Z.* 1911, **32**, pp. 1001–1002

9 WALSH, R.: 'Inductive ratio arms in alternating current bridge circuits', *Philos. Mag.*, 1930, **10**, pp. 49–70

10 STARR, A.T.: 'A note on impedance measurement', *Wireless Engineer & Experimental Wireless*, 1932, **9**, pp. 615–617

11 KIRKE, H.L.: 'Radio frequency bridges', *J. IEE*, 1945, **92**, Pt. III, pp. 2–7

12 YOUNG, C.H.: 'Measuring inter-electrode capacitances', *Bell Lab. Rec.*, 1946, **24**, pp. 433–438

13 CLARK, H.A.M.: and VANDERLYN, P.B.: 'Double ratio a.c. bridges with inductively coupled ratio arms', *Proc. IEE*, 1949, **96**, Pt. III, pp. 189–202

14 WILHELM, H.T.: 'Impedance bridges for the megacycle range', *BSTJ*, 1952, **31**, pp. 999–1012

15 OATLEY, C.W., and YATES, J.G.: 'Bridges with coupled inductive ratio arms as precision instruments for the comparison of laboratory standards of resistance or capacitance', *Proc. IEE*, 1954, **101**, Pt. III, pp. 91–100

16 PINCKNEY, C.B.: 'A method for calibration of precision voltage dividers', *AIEE Trans.*, 1959, **78**, pp. 182–185

17 HILL, J.J.: and MILLER, A.P.: 'A seven-decade adjustable-ratio inductively-coupled voltage divider with 0·1 part per million accuracy', *Proc. IEE*, 1962, **109B**, pp. 157–162

18 ZAPF, T.L.: 'The calibration of inductive voltage dividers and analysis of their operational characteristics', *ISA Trans.*, 1963, **2**, pp. 195–201

19 ZAPF, T.L., *et al.*: 'Inductive voltage dividers with calculable relative corrections', *IEEE Trans.*, 1963, **IM-12**, pp. 80–85

20 BINNIE, A.J.: 'An inductive decade divider of low output impedance', *J. Sci. Instrum.*, 1964, **41**, pp. 747–750

21 SZE, W.C., *et al.*, 'An international comparison of inductive voltage divider calibrations at 400 and 1000 Hz', *IEEE Trans.*, 1965, **IM-14**, pp. 124–131

22 HOER, C.A., and SMITH, W.L.: 'A 2:1 inductive voltage divider with less than 0·1 ppm error to 1 MHz', *J. Res. NBS*, 1967, **71C**, pp. 101–109

23 HILL, J.J., and DEACON, T.A.: 'Theory, design and measurement of inductive voltage dividers', *Proc. IEE*, 1968, **115**, pp. 727–735

24 DEACON, T.A., and HILL, J.J.: 'Two-stage inductive voltage dividers', *Proc. IEE*, 1968, **115**, pp. 888–892

25 HILL, J.J., and DEACON, T.A.: 'Voltage ratio measurement with a precision of parts in 10^9 and performance of inductive voltage dividers', *IEEE Trans.*, 1968, **IM-17**, pp. 269–278

26 HOER, C.A., and SMITH, W.L.: 'A 1 MHz binary inductive voltage divider with ratios of 2^n to 1 or $6n$ dB', *IEEE Trans.*, 1968, **IM-17**, pp. 278–284

27 DEACON, T.A.: 'Accurate (2 parts in 10^7) inductive voltage divider for 20–200 Hz', *Proc. IEE*, 1970, **117**, pp. 634–640

28 HILL, J.J.: 'An optimised design for a low-frequency inductive voltage divider', *IEEE Trans.*, 1972, **IM-21**, pp. 368–372

29 BOOTHBY, O.L., and BOZORTH, R.M.: 'A new magnetic material of high permeability', *J. Appl. Phys.*, 1947, **18**, pp. 173–176

30 BARDELL, P.R.: 'Magnetic materials in the electrical industry' (Macdonald, 1960, 2nd edn.), chap. 4

31 COTTON, H.: 'Electrical technology' (Pitman, 1957, 7th edn.), chap. 17

32 GOLDING, E.W., and WIDDIS, F.C.: 'Electrical measurements and measuring systems' (Pitman, 1963, 5th edn.), chaps. 5, 9 and 14

33 WELSBY, V.G.: 'The theory and design of inductance coils' (Macdonald, 1960, 2nd edn.), chap. 5

34 HAMMOND, P.: 'Electromagnetism for engineers' (Pergamon, 1964), chap. 7

35 M.I.T.: 'Magnetic circuits and transformers' (Wiley, 1943), chap. 5

36 TERMAN, F.E.: 'Radio engineers handbook' (McGraw Hill, 1943), p. 31

37 McLACHLAN, N.W.: 'Bessel functions for engineers' (Oxford University Press, 1934), chap. 9

38 HARNETT, D.E., and CASE, N.P.: 'The design and testing of multi range receivers', *Proc. IRE*, 1935, **23**, pp. 578–593

39 GAINSBOROUGH, G.F.: 'A method of calibrating standard-signal generators and radio frequency attenuators', *J. IEE*, 1947, **94**, Pt. III, pp. 203–210

40 GRANTHAM, R.E., and FREEMAN, J.J.: 'A standard of attenuation for microwave measurements', *Trans. AIEE*, 1948, **67**, pp. 329–335

41 WEINSCHEL, B.O., SORGER, G.U., and HEDRICH, A.L.: 'Relative voltmeter for v.h.f/u.h.f. signal generator calibration', *IRE Trans.*, 1959, **I-8**, pp. 22–31

42 COOK, C.C., and ALLRED, C.M.: 'An excitation system for piston attenuators', *IEEE Trans.*, 1971, **IM-20**, pp. 10–16

43 ALLRED, C.M., and COOK, C.C.: 'A precision r.f. attenuation calibration system', *IRE Trans.*, 1960, **I-9**, pp. 268–274

44 GRIESHEIMER, R.N.: 'Cut-off attenuators' *in* MONTGOMERY, C.G. (Ed.): 'Technique of Microwave measurements' (McGraw Hill, 1947), chap. 11

45 BROWN, J.: 'Corrections to the attenuation constants of piston attenuators', *Proc. IEE*, 1949, **96**, Pt. 3, pp. 491–495

46 BARLOW, H.M., and CULLEN, A.L.: 'Microwave measurements' (Constable, 1950), pp. 242–253

47 WEINSCHEL, B.O., and SORGER, G.U.: 'Waveguide below cut-off attenuation standard', *Acta Imeco*, 1964, pp. 407–423

48 YELL, R.W.: 'Development of a high precision waveguide beyond cut-off attenuator', *CPEM digest*, 1972, pp. 108–110

49 HOLLWAY, D.L., and KELLY, F.P.: 'A standard attenuator and the precise measurement of attenuation', *IEEE Trans.*, 1964, **IM-13**, pp. 33–44

Chapter 4

Microwave attenuation standards

Many different types of coaxial and waveguide variable attenuators have been described in the literature.[1-3] However, only two of them, the waveguide rotary vane attenuator and the microwave piston attenuator, are really suitable for use as the attenuation reference standard in the r.f. substitution method of attenuation measurement, which is described in Chapter 6. Thus, we shall only discuss these two types of variable attenuators here.

4.1 Rotary vane attenuators

The rotary vane attenuator[4-22] was invented simultaneously by E.A.N. Whitehead[*] and A.E. Bowen[†], and the first description of it was given by Southworth.[4] The essential parts of a rotary vane attenuator are shown in Fig. 4.1. The two end vanes are fixed in a direction perpendicular to the incident electric vector and the attenuation is varied by rotating the central section.

Let the incident wave have an amplitude E and let the central vane be at an angle θ relative to the end vanes. At the input end of the central section, let the incident wave be resolved into two components: $E \cos \theta$ perpendicular to the central vane and $E \sin \theta$ parallel to it (see Fig. 4.2). If there is sufficient attenuation in the central vane, the $E \sin \theta$ component will be almost completely absorbed, and the $E \cos \theta$ component will emerge almost unchanged. It is now necessary to resolve this emergent wave into two further components: $E \cos^2 \theta$ perpendicular to the end vanes and $E \sin \theta \cos \theta$ parallel to them (see Fig. 4.2). The vane at the far end will then absorb this latter component and the $E \cos^2 \theta$ component will be transmitted with very

---

[*]Elliott Brothers (London) Ltd.
[†]Bell Telephone Laboratories.

little loss to the output. Thus, the attenuation A in dB that is produced at this angle θ is given by

$$A = 20 \log_{10} \frac{E}{E \cos^2 \theta} + A_0 = 40 \log_{10} (\sec \theta) + A_0 \quad (4.1)$$

where A_0 is the residual attenuation when all three vanes lie in the same plane. Thus, the incremental attenuation that is provided when the central vane is moved from 0 to θ is $(A - A_0)$.

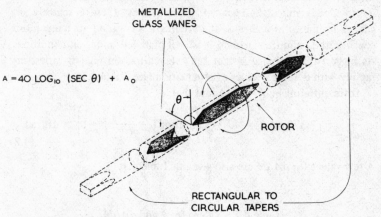

Fig. 4.1 Basic parts of a rotary attenuator

$A - A_0$, dB	0	3	10	20	30	40	50	60
θ, deg	0·000	32·712	55·782	71·565	79·757	84·261	86·776	88·188

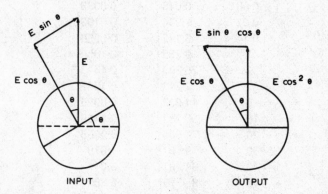

Fig. 4.2 Vectors at the input and output of the circular section

The fixed vane is needed at the input end to make the unit bilateral. Tapers are needed on the ends of all of the vanes to minimise reflections.

It is desirable to use choke-type rotating joints at each end of the rotor and an ABAC for designing them is available in the literature.[23] When the highest accuracy is sought, very lossy material should be placed around the outsides of the chokes.[17] A worm drive is often used to rotate the central section, and a drum or scale, graduated directly in decibels, in accordance with eqn. 4.1, is usually attached to the worm input shaft.

The rotary attenuator has many advantages over the other types of variable waveguide attenuators which are still in use; the attenuation it provides is almost independent of frequency, there is scarcely any change in phase through it as the attenuation is varied, its temperature coefficient is almost zero and it is well matched under all conditions. Its biggest disadvantage is that the attenuation through it varies very rapidly with θ at high values of attenuation (see Table 4.1).

Differentiating eqn. 4.1, we find that

$$\frac{dA}{d\theta} = 17 \cdot 372 \tan \theta = 17 \cdot 372 \left\{ \text{antilog} \left(\frac{A - A_0}{20} \right) - 1 \right\}^{1/2} \text{ dB/rad} \tag{4.2}$$

A few values for $dA/d\theta$ are also given in Table 4.1.

Table 4.1   Values for θ and $dA/d\theta$

$A - A_0$ dB	θ degrees	$\dfrac{dA}{d\theta}$ dB/degree
0·001	0·615	0·0033
0·01	1·944	0·0103
0·1	6·142	0·0326
1	19·255	0·1059
2	26·969	0·1543
3	32·712	0·1947
6	44·932	0·3025
10	55·782	0·4458
20	71·565	0·9095
30	79·757	1·678
40	84·261	3·017
50	86·776	5·383
60	88·188	9·583
70	88·981	17·046
80	89·427	30·316

From this table it can be seen that, in order to achieve a resetting accuracy not worse than ± 0·01 dB up to 60 dB, the rotor angle must be set to an accuracy of approximately ± 0·001°. A general family of curves based on eqn. 4.2 is presented in Fig. 4.3. Eqn. 4.2 can also be used to determine the errors that occur if the graduated drum or scale is incorrectly aligned relative to the rotor. This is, in fact, a very common and important source of error in a rotary attenuator. However, it can be corrected by slightly changing the angular position of the drum or scale on the input shaft. Other sources of error in a rotary attenuator are: end vane misalignment, insufficient attenuation in the central vane, internal reflections, eccentricity in the worm drive and external leakage through the rotating joints. All of these sources of error will now be discussed. Most of them have been analysed in the literature by Hand,[5] Mariner,[7] James,[8] Larson[10, 12, 14, 22] and Otoshi and Stelzried.[18]

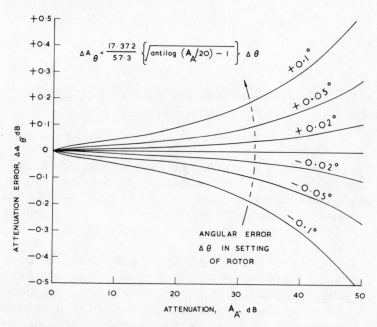

Fig. 4.3 Resettability error with a rotary attenuator

When the two end vanes are misaligned and the total angle between them is 2ϵ (see Fig. 4.4), the departure from the $40 \log_{10} (\sec \theta)$ law is given in dB by

$$\Delta A_\epsilon = 20 \log_{10} \frac{\cos^2 \theta \cos^2 \epsilon}{\cos(\theta - \epsilon)\cos(\theta + \epsilon)} \qquad (4.3)$$

$$\approx 8 \cdot 686 \left(\frac{\epsilon}{57 \cdot 3}\right)^2 \{10^{(A_A/20)} - 1\} \qquad (4.4)$$

where $A_A = A - A_0$, and ϵ is in degrees.

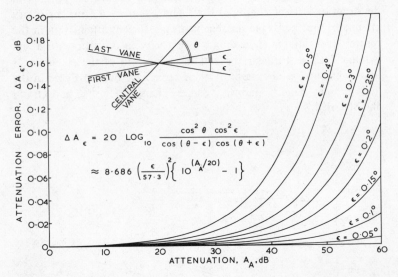

Fig. 4.4 End vane misalignment error

End vane misalignment always increases the attenuation, and the family of curves given in Fig. 4.4 shows that ΔA_ϵ can become very large at high values of attenuation unless great care is taken to align the two end vanes very accurately. To keep this source of error below $0 \cdot 01 \, \text{dB}$ up to 60 dB, the upper limit on ϵ is $0 \cdot 06°$, which corresponds to a total misalignment of $0 \cdot 12°$.

Ideally, one would like to have infinite attenuation in the central vane but, in practice, it is difficult to obtain vanes which will give an attenuation greater than about 100 dB. Let the maximum attenuation provided by the central vane be denoted by A_V. Then, at an incremental attenuation value of A_A, the departure from the $40 \log_{10}(\sec \theta)$ law, due to insufficient attenuation in the central vane, has upper and lower limits which are given in dB by

$$\Delta A_V = 20 \log_{10} \frac{10^{A_V/20}}{10^{A_V/20} \pm \{10^{A_A/20} - 1\}} \qquad (4.5)$$

$$\approx \pm 8.686 \frac{10^{A_A/20} - 1}{10^{A_V/20}} \qquad (4.6)$$

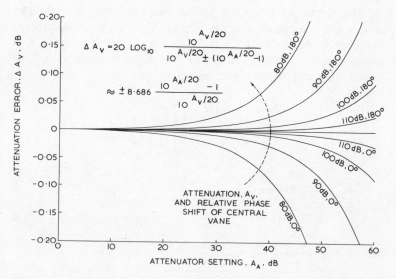

Fig. 4.5 Error due to insufficient attenuation in central vane

Fig. 4.5 gives curves of ΔA_V versus A_A for several different values of A_V. The glass or mica plate on which the central metal film is deposited causes the phase constants to be different for the signal components that are perpendicular and parallel to the vane. Thus, the incompletely absorbed component can have any phase relationship relative to the $E\cos^2\theta$ component. Only the curves corresponding to in-phase and anti-phase conditions are shown in Fig. 4.5 but other phase angles simply give error curves that lie between the ones shown in this diagram.

From Fig. 4.5, it can be seen that a 100 dB central vane can produce an error of ± 0.087 dB at 60 dB. Thus, when designing a rotary vane attenuator, it is extremely important to make the attenuation in the central vane as high as possible.

In a rotary vane attenuator, there are 10 main places where reflections can occur, namely, at both ends of the three vanes, at both of the rotating joints and in each of the rectangular to circular tapers. Thus , there are $9 + 8 + 7 + 6 + 5 + 4 + 3 + 2 + 1 = 45$ different routes by which wave components can reach the output after being reflected twice. The wave components that are reflected an odd number of times are returned to the source and those that are reflected four times should be negligibly small. Hence, a complete theoretical treatment of internal reflections in a rotary attenuator would be very complex. However, Hand[5] has tackled this problem by assuming that the rotating joints and tapers are perfect and considering only the two double reflection paths shown in Fig. 4.6. Then, by assuming that the components perpendicular to the vanes are not reflected at all, it is found that the deviation from the theoretical law due to internal reflections has upper and lower limits given by

$$\Delta A_R = 20 \log_{10} \frac{1}{1 \pm 2\Gamma^2 \sin^2 \theta} \qquad (4.7)$$

$$\approx \pm 17{\cdot}372\, \Gamma^2 \left\{1 - 10^{-A_A/20}\right\} \qquad (4.8)$$

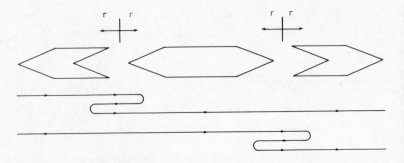

Fig. 4.6 Two double reflection paths used when deriving eqn. 4.7

where Γ denotes the voltage reflection coefficient at each of the 4 places in Fig. 4.6 where reflections occur.

Fig. 4.7 contains a general family of curves based on eqn. 4.7 and, unlike the other errors that can occur in a rotary attenuator, it is seen that the internal reflection error can become quite serious at low values of attenuation. Thus, the tapers on the ends of the vanes should be designed very carefully[1] to keep the value of Γ as low as possible.

In many commercially available rotary attenuators, the effect of eccentricity in the worm drive shows up very clearly in the attenuation error curve; so this will be discussed next.

When the worm eccentricity gives a peak angular error ψ, with a starting phase ϕ and n complete turns are required to move the central vane through $90°$, the angular error at an angle θ is given by

$$\Delta \theta_w = \psi \sin (4n\theta + \phi) \qquad (4.9)$$

Combining eqn. 4.9 with eqn. 4.2, the deviation from the theoretical law due to worm eccentricity is seen to be

$$\Delta A_w = 17 \cdot 372 \; \psi \tan \theta \sin (4n\theta + \phi) \tag{4.10}$$

where

$$\theta = \sec^{-1} \{\text{antilog} \, (A_A/40)\} \tag{4.11}$$

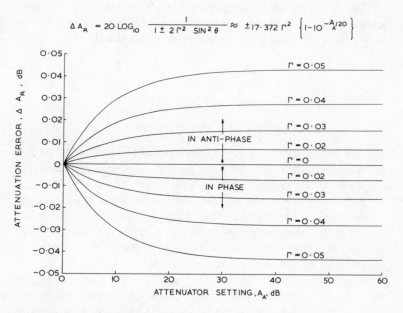

Fig. 4.7 Error due to internal reflections

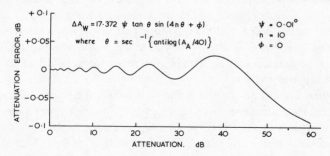

Fig. 4.8 Error curve caused by eccentricity in the worm drive of a rotary attenuator

Fig. 4.8 shows how ΔA_w varies with A_A when $\psi = 0 \cdot 01°$, $n = 10$ and $\phi = 0$. This error curve is so different from all of the others that worm eccentricity is quite unmistakable. The error caused by external leakage between the two rotating joints has already been analysed (see eqn. 2.83 and Fig. 2.9). In practice, several different sources of error are likely to be present simultaneously and the appropriate error expressions must then be combined with each other to find the resultant error curve.

When a rotary vane attenuator suffers from both end vane misalignment and insufficient attenuation in the central vane, it is shown in Appendix 4 that the attenuation is given approximately by

$$A'' = 20 \log_{10} \frac{1}{\{M^2 + N^2 + 2MN \cos \xi\}^{1/2}} \qquad (4.12)$$

where

$$M = (\cos^2 \theta - \epsilon^2 \sin^2 \theta) \qquad (4.13)$$

$$N = (\sin^2 \theta - \epsilon^2 \cos^2 \theta) \, 10^{-A_V/20} \qquad (4.14)$$

and ξ is the differential phase change that occurs in the central section between the components that are parallel and perpendicular to the vane. Fig. 4.9 shows how A'' varies with θ over the range $80°$ to $100°$ for various values of ϵ and A_V. Double humped curves of this type are frequently found in practice. Thus, errors will occur throughout the whole range if the central vane is rotated to the first maximum attenuation position and the scale is then set to read $90°$. The correct procedure is to set the scale so that equal values of attenuation are obtained at angles which are displaced by equal amounts below and above $90°$. Suitable angles for this purpose are $80°$ and $100°$ as they give quite large values for $\pm dA''/d\theta$ and also a high signal to noise ratio on a wide dynamic range attenuation-measuring equipment.

Rotary attenuators are available from many commercial firms and a typical accuracy specification on a mass-produced one is, " $\pm 2\%$ of reading in dB or 0.1 dB, whichever is greater". If attenuation measurements are required only to this accuracy, a commercially available rotary attenuator can be used as the reference standard in an r.f. substitution equipment.

When the highest possible accuracy is required, commercially available rotary attenuators become inadequate. On one popular type, the scale length is so short that the resetting accuracy is limited to $\pm 0.05°$. All of the commercially available instruments that are fitted with worm drives have a significant amount of backlash. In one brand-new rotary attenuator that was examined at RRE, the backlash was found to be $0.028°$. Several weeks were spent at RRE trying to eliminate the backlash from an American X-band rotary attenuator that had been fitted with a precision gearbox and three readout dials, but it could not be reduced below 0·003°. In this particular rotary attenuator, approaches in the same direction through different angles to a specified setting on the dials produced errors as high as 0·016° in the angular setting of the central vane. Both of these troubles were traced to the worm drive. The worm eccentricity error discussed earlier is a familiar shortcoming of commercially-available rotary attenuators, and, in many of them, a serious change in the calibration is produced by an impact with one of the end stops.

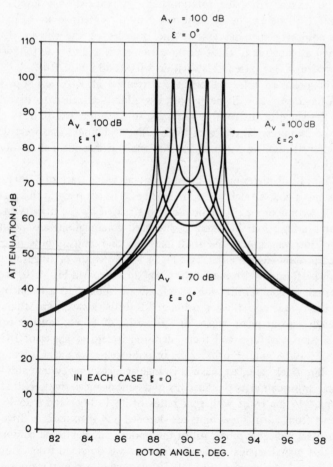

Fig. 4.9 Attenuation versus rotor angle
With both end vane misalignment and insufficient loss in the central vane

To overcome these troubles, rotary attenuators with optical angular measuring systems have been developed recently at NBS by Little, Larson and Kinder[17] and at RRE by Warner, Watton and Herman.[21]

In the NBS design, an engraved glass scale is fitted to the rotor and it is observed with a dual microscope that has two apertures 180° apart to reduce the eccentricity error. The scale has an absolute accuracy of ± 1 second of arc and the resolution is also ± 1 second of arc, so the total uncertainty about the angular setting of the central vane is ± 2 seconds of arc (i.e. ± 0·00056°). The worm gear is attached to the rotor by a bellows drive. A very strong housing is used and the rotor is mounted in finest quality triple A bearings.

The leakage from the rotating joints is reduced to a level more than 120 dB below the incident power by the extensive use of microwave-absorbing material around the outsides of the chokes. Three X-band attenuators of this type have been made and they all follow the theoretical $40 \log_{10} (\sec \theta)$ law within ± 0·002 dB up to 20 dB.

The precision rotary attenuator developed at RRE also operates at X-band. An $11''$ diameter optical grating is attached to its rotor, and the angle of the rotor is displayed to a resolution of 0·001° on a row of six numerical indicator tubes, using techniques that were developed by Russell in 1966 at the National Engineering Laboratory in Scotland.[24]

The $11''$ diameter optical grating contains a fine track of 3600 radial lines and a coarse track of 360 radial lines. The accuracy of this grating is ± 1 second or arc. To minimise errors caused by eccentricity in the rotor bearings, four light-source photocell combinations are mounted at 90° intervals around the 3600 line track and their outputs are combined in summing amplifiers (see Fig. 4.10). When the grating is rotated two waveforms, which are labelled 'sin' and '−cos' in Fig. 4.10, and an inverted version of the sine waveform are applied to a chain of 50 precision resistors, thus providing 48 further waveforms that rise through zero at phase angles of 3·6°, 7·2°, 10·8° etc, relative to the initial sine waveform and then fall through zero at angles of 183·6°, 187·2°, 190·8° etc. Further pairs of photocells give outputs from the 360 line track, and, to make this angular measuring system absolute, coarse information is provided by three mechanical digitisers that are coupled to the rotor with gear ratios of 36:1, 3·6:1 and 0·36:1. The signals from all of these units are decoded and then used to drive the numerical indicator tubes. All of the optical and electronic equipment for this digital angular measuring system was obtained from Whitwell Electronic Developments Ltd. of Glasgow, who market systems similar to the one devised by Russell.[24] This equipment includes a number of

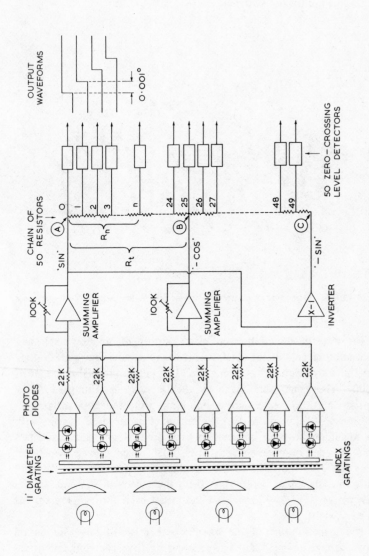

Fig. 4.10 Simplified block diagram of the early stages of the absolute digital angular measuring system

lag-lead circuits which cause all changes in the displayed digits to be synchronised to the change-over points associated with the fine optical track. The datum can be set to any required value by adjustment of six 10-pole multiwafer switches.

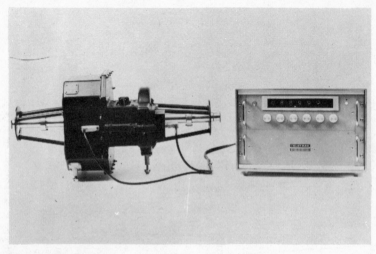

Fig. 4.11 RRE *X*-band rotary attenuation standard with its associated electronic box

The waveguide sections were electroformed on stainless steel mandrels and then gold plated. The rectangular to circular tapers are 5″ long and the rotor section has a length of 12″ and an internal diameter of 1·000″. The vanes were made by the Decca Radar Company. They consist of 200 Ω per square metal films protected with magnesium fluoride on 0·015″ thick glass plates and they are used in pairs with the metallised surfaces adjacent to each other along diameters. These pairs of vanes are precisely located in 0·030″ deep grooves, which run along the inside surfaces of the circular waveguide sections. The end vanes are aligned to better than 0·1°, the tapers on the ends of the glass plates are 3·2–4″ long and the attenuation through the central vane is 104 dB.

The main housing of the attenuator consists of three aluminium alloy castings designed for lightness and rigidity. The complete attenuator, together with its associated electronic box, is shown in Fig. 4.11. The casting appearing on the left can be easily removed to give access to the optical grating and the four reading heads which are shown in Fig. 4.12. The rotor is mounted in low radial runout angular contact bearings with ABEC7 tolerances. The manual drive gearbox gives a ratio of 500:1 with the rotor. There are no end stops in this attenuator and all four quadrants are available for use.

Fig. 4.12 View of RRE rotary attenuation standard
End cover removed, showing optical grating and four reading heads

According to measurements made at 10 GHz against the British National Standard, (which is described in Chapter 8) this rotary attenuator follows the $40 \log_{10} (\sec \theta)$ law to within ± 0.0015 dB up to 16 dB (see Table 4.2). To check the resettability of this rotary attenuator, it was set 50 different times to an angle of 88·188°, corresponding to a nominal attenuation of 60 dB, and the 50 measured values for the attenuation lay between 60·036 dB and 60·044 dB. This total spread of 0·008 dB corresponds to a resetting accuracy of better than 0·001°.

At 10 GHz, this attenuator has a residual attenuation of 0·15 dB and a v.s.w.r. that varies from 1·045 at 0 dB to 1·016 at 60 dB. Over the range 0 to 60 dB the phase of the signal transmitted through it does not change by more than 1°. This rotary attenuator was developed for use as a primary attenuation standard. Several less-expensive precision rotary attenuators have been developed at RRE for use as transfer standards and these are described in Chapter 13.

Table 4.2 **Attenuation results at 10 GHz on the RRE rotary atten-uator with an absolute digital angular measuring system**

Nominal value	Average measured value on modulated subcarrier system	Deviation from theoretical value
dB	dB	dB
1	0·9992	−0·0008
2	1·9987	−0·0013
3	2·9987	−0·0013
4	3·9989	−0·0011
5	4·9989	−0·0011
6	6·0000	0·0000
7	7·0003	0·0003
8	8·0006	0·0006
9	9·0004	0·0004
10	10·0002	0·0002
11	11·0008	0·0008
12	12·0003	0·0003
13	13·0007	0·0007
14	14·0003	0·0003
15	15·0014	0·0014
16	16·0005	0·0005

A precision rotary attenuator is an excellent device for producing an attenuation change of 0·001 dB to check the short-term jitter and long-term drift of a microwave attenuation measuring equipment.

4.2 Microwave piston attenuators

4.2.1 Introduction

The use of piston attenuators at intermediate frequencies, e.g. 60 MHz has been described in Chapter 3. These devices have also been used at microwave frequencies because of their calculable attenuation rate, large dynamic range and uniform scale of attenuation. In particular they found a use in signal generators designed to measure the noise factor of microwave receivers. However, stable gas discharge tubes, capable of providing reliable sources of microwave noise, superseded the microwave signal generator and there have been few examples of designs of microwave piston attenuators during recent years. Although the basic theory is identical with that given in Chapter 3 and Appendix 3, the relative importance of various design considerations differs in the two applications because of the difference in the ratio of wavelength to geometric dimensions in practicable designs.

From eqns. 3.54 and 3.55 it follows that the attenuation rate for the H_{11} mode in a circular waveguide beyond cut-off is given by

$$R_{pa} = 54 \cdot 575 \left\{ \left(\frac{\rho_{11}}{\pi d} \right)^2 - \left(\frac{f}{c} \right)^2 \right\}^{1/2} \text{ dB/mm} \qquad (4.15)$$

Assuming the diameter $d = 30$ mm for a 60 MHz attenuator, the two terms within the square root are $3 \cdot 8164 \times 10^{-4}$ and $4 \cdot 0055 \times 10^{-8}$. Because the first term predominates, the attenuation rate is almost independent of frequency and has a value of $1 \cdot 066$ dB/mm. This is a sensible compromise between the needs for an open scale to provide adequate discrimination and for a compact traverse capable of providing a useful dynamic range of at least 100 dB. By differentiating R_{pa} with respect to frequency, a frequency sensitivity of about $1 \cdot 86 \times 10^{-12}$ dB/mm/Hz is obtained, so that a 17% frequency change (i.e. 10 MHz in 60 MHz) alters the attenuation rate by only $0 \cdot 000018$ dB/mm. It is pointed out in Section 3.3.4 that the effective diameter to be used in eqn. 4.15 differs from the physical diameter because of penetration of the electromagnetic field into the metal, i.e. the skin effect. At 60 MHz the skin depth in brass is about $0 \cdot 017$ mm.

Any attempt to scale the diameter of the attenuator in proportion to wavelength would result in an impracticably small diameter of $0 \cdot 2$ mm for a frequency of 9 GHz. Moreover the attenuation rate of approximately 160 dB/mm would make difficult the discrimination of small changes in attenuation. The absolute frequency sensitivity would be the same as for the 60 MHz attenuator. A more practical design approach is to use a bore of about 10 mm diameter to ease the mechanical difficulties. Now the two terms under the square root sign of eqn. 4.15 are more nearly comparable, i.e. $34 \cdot 375 \times 10^{-4}$ and $9 \cdot 0125 \times 10^{-4}$ and the frequency sensitivity is $9 \cdot 31 \times 10^{-11}$ dB/mm/Hz. Thus a 17% frequency change, i.e. 150 MHz, alters the attenuation rate by $0 \cdot 017$ dB/mm, which is significant with respect to the calculated rate of $2 \cdot 75$ dB/mm at 9 GHz. The skin depth in brass at 9 GHz is smaller in inverse proportion to the square root of the frequency, i.e. it is about $0 \cdot 0014$ mm. Such an attenuator would provide a dynamic range of 150 dB for a 55 mm traverse and a movement of 1 μm along the cylinder would produce an attenuation charge of $0 \cdot 00275$ dB. Allowing for input and output couplings together with a non-linear region of attenuation, the overall length of the attenuator barrel will be about ten times the diameter which is a reasonable aspect ratio for accurate manufacture. Means must be provided for moving the output section, the 'piston', with respect to the input section in a precise manner and any microwave leakage paths between input and output must be minimised. These are both difficult requirements and to some degree interact, because a loose fitting piston within the attenuator barrel will encourage leakage. It should be pointed out that, for precise measurements, the leakage signal should be some 80 dB down on the direct signal, i.e. leakage paths between input and output should be 200 dB or more. The relative movement of input and output requires a flexible coupling at one end at least, unless the source or detector can be moved without impairing the precise motion of the piston within the barrel. Such flexible couplings can take the form of totally shielded coaxial lines or sliding waveguide sections. The loss through either type of flexible coupling may vary with position and thus provide an increased uncertainty in measurement.

4.2.2 Representative piston attenuator designs

Some designs of piston attenuators for 3 GHz, 9 GHz and 24 GHz which originated during World War 2 are described by R.H. Griesheimer.[25] The TPS-15 attenuator for 3 GHz used a 12·7 mm diameter barrel in which slid a close fitting tube with about 10 fingers cut in the end to improve the electrical contact. Both input and output were in 50 Ω coaxial line, the inner being bent at the end and grounded to the outer, thus forming input and output coupling loops (Fig. 4.13). The piston was driven along the barrel by a rack and pinion, using a large diameter steel rod to improve guidance. The piston position was indicated by a dial attached to the pinion gear with a consequential backlash error. The change in field pattern between the TEM coaxial mode and the H_{11} mode in the waveguide, combined with the simple form of coupling, results in many modes being launched (the geometry was too small to permit sophisticated mode filters). Consequently, the attenuator law departed from linearity by 0·5 dB for an insertion loss of 30 dB. The attenuator itself had no input or output matching devices so that about 16 dB of further attenuation at both input and output was required to present a v.s.w.r. of about 1·05 to source and load. In use, the piston fingers scored the barrel and there was appreciable leakage of microwave power between the piston and the barrel.

Attempts were made to improve the mode purity in other designs by the replacement of the coupling loop by an iris in the side wall of the coaxial line, the waveguide beyond cut-off being orthogonal to the coaxial line. These were successful in reducing the insertion loss (exclusive of lossy cable) to 20 dB for 0·5 dB departure from linearity. A more satisfactory design, the model *S* attenuator Fig. 4.14, incorporated a cavity between the input coaxial loop and the input iris of the waveguide beyond cut-off, which was also of 12·7 mm diameter. The output coupling loop was enlarged and included a 50 Ω resistive disc for impedance matching, which produced a v.s.w.r. of less than 1·15. Two resistive discs were spaced a quarter wavelength apart in the input coaxial line reducing the v.s.w.r. to 1·35 when the cavity was on resonance. These discs introduced about 6 dB of loss but removed the need for the lossy cable. A pair of helical springs, spaced a quarter of a wavelength apart were used between the piston and the barrel to minimise leakage without introducing severe mechanical problems of alignment between barrel, piston and external guide rods. A depth gauge associated with the piston eliminated the backlash problems of a rack and pinion with the dial on the pinion shaft. With this design of attenuator the performance was within 0·5 dB of linearity for an insertion loss (including the piston matching disc) of 6 dB, and within 0·1 dB of linearity at 14 dB overall attenuation.

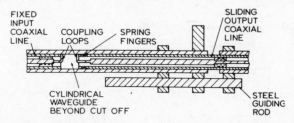

Fig. 4.13 Diagram of TPS-15 attenuator (3 GHz)

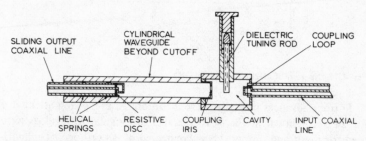

Fig. 4.14 Diagram of Model S attenuator (3 GHz)

A significant departure in techniques was introduced in a 9 GHz attenuator, Fig. 4.15, by feeding the circular waveguide beyond cut-off, of 8·5 mm diameter, directly from an iris in the end wall of a standard X-band rectangular waveguide. The iris extended across the diameter of the cylindrical guide and had its long axis parallel to the broad dimension of the rectangular waveguide. The piston used a large rectangular loop to feed a coaxial line. A bushing on the outside of the piston made sliding contact with the attenuator barrel and half-wavelength chokes in front of the bushing minimised leakage. About 20 dB attenuation was needed to bring the values within 0·5 dB of linearity.

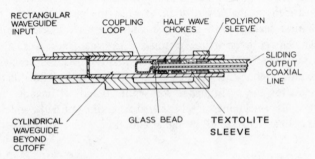

Fig. 4.15 Diagram of attenuator for 9 GHz

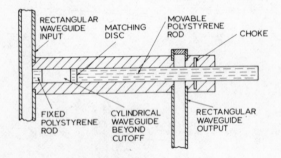

Fig. 4.16 Diagram of attenuator for 24 GHz

Further advances in design were incorporated in a piston attenuator for 24 GHz, shown in Fig. 4.16. Parallel rectangular waveguides were used for input and output, being connected by an orthogonal cylindrical tube. The diameter of this tube was chosen to be beyond cut-off when air filled but to propagate the H_{11} mode (and no others) when filled with a polystyrene rod. A small length of this rod, approximately a half guide-wavelength long, acted as a mode filter for power coupled out of the input guide and, together with a short circuit appropriately placed in the rectangular guide, provided a good impedance match between the rectangular guide and the dielectric-filled tube. The piston took the form of a long polystyrene rod fitted with a resistive matching disc. A choke in the cylindrical barrel behind the junction with the rectangular waveguide prevented microwave leakage and ensured good coupling into the output guide. The polystyrene rod was driven by a micrometer. The design had several advantages in that the input and output guides were fixed relative to each other, the use of a dielectric rod minimised wear on the barrel and eliminated variable contact troubles associated with previous designs. Moreover, the rods could be brought into close contact and the attenuation could be predicted even in the non-linear region of attenuation. However, there were some mechanical problems associated with the resistive matching discs and care had to be taken to avoid any lost movement between the micrometer drive and the dielectric rod.

The attenuator shown in Fig. 4.17 was used in the TRE Signal Generator Type 47 for 3·3 GHz and was described by L.G.H. Huxley.[26] As in the TPS-15 attenuator, the sliding output coaxial line was driven by a rack and pinion. A bolometer filament was used as a launching loop, thus enabling the input power to be monitored by the filament glow.

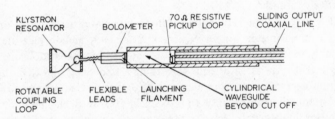

Fig. 4.17 Diagram of attenuator for TRE Signal Generator Type 47 (3·3 GHz)

A.C. Gordon-Smith[27] described an attenuator working at 48·2 GHz, which used a dielectric rod as a pickup device in an arrangement somewhat similar to that of Fig. 4.16. However, in Gordon-Smith's design, the rod was fixed with respect to the output guide which moved as a whole relative to the input guide with a consequent flexing of the output guide. The attenuation rate was 0·71 dB/mm as compared with an estimated value of 0·72 ± 0·01 dB/mm and the attenuator was sensibly linear for attenuation values of 10 dB or more.

An attenuator for 3·3 GHz, using coaxial input and output lines but incorporating an ebonite rod as pickup element was described by J. Munier.[28] In this design, the length of the ebonite rod was chosen to be resonant at the operating frequency and a miniature resistor formed part of the input coaxial line near the coupling loop. The attenuation rate was stated to be 1·1905 dB/mm.

The designs described so far have used the H_{11} mode, but the possibility of using an E mode is discussed by Griesheimer.[25] A serious disadvantage is that, in the waveguide beyond cut-off, the attenuation rate for the E_{01} mode is greater than that for the H_{11} mode. Therefore, the relative strength of any spurious H_{11} mode signal in an E_{01} mode attenuator will increase as the attenuation of the E_{01} mode signal increases. A.B. Giordano[29] described an E_{01} mode attenuator for use at 3 GHz, with coaxial input and output lines, but using axial rods rather than radial loops to excite and pick up the evanescent wave.

4.2.3 Piston attenuator as an absolute standard

Although microwave piston attenuator design was not actively pursued during the period 1950 to 1970, the device still has attractive features in that it is one of the few instruments which enables a microwave parameter to be established in terms of the base units of length and time. The realisation of this concept depends directly on the solution of the difficult mechanical problems by use of modern techniques and on the incorporation of microwave techniques such as ferrite isolators and frequency controlled sources which were not developed until relatively recently.

The first problem is to manufacture an attenuator barrel of the order of 10 mm diameter for an H_{11} mode attenuator and to measure accurately the absolute value of this diameter and its variation along the length. Differentiation of eqn. 4.15 using a frequency of 9 GHz shows that an uncertainty in bore diameter of 0·75 μm is required to restrict the attenuation uncertainty to 0·01 dB in 100 dB. Such fine limits on diameter immediately call for temperature control of the attenuator barrel. Taking the expansion coefficient of brass for example as 19 $\times$ 10^{-6} per °C a temperature change of 1°C causes a diameter change of 0·19 μm and an attenuation change of about 0·0025 dB in 100 dB. A further problem is that the diameter required to calculate the attenuation rate must allow for the skin effect. As noted in Chapter 3, the microwave conductivity differs from the d.c. value and is dependent on the machining history and surface roughness of the metal. Therefore, any estimate of the skin depth δ may be in error because of the uncertainty in the value of conductivity to use. If a value of 15 $\times$ 10^6 siemens/m is assumed for brass at 9 GHz, the skin depth is 1·4 $\times$ 10^{-3} mm, i.e. 1·4 μm. J. Brown[30] shows that, when the bore diameter is a small fraction of the wavelength, the relative change in attenuation rate approaches an asymptotic value of δ/d, and that, if the second term in the square root of eqn. 4.15 can be neglected, this is equivalent to replacing the physical diameter d by an electrical diameter $(d + \delta)$. However, in the microwave region neither of these approximations are valid and any specific example must be evaluated for the relevant wavelength and diameter. Nevertheless, the physical concept remains that the effective diameter has a value of $(d + x\delta)$, where x is a number of the order of unity. Assuming that x has a value between 2 and 3 for a particular geometry, the skin effect produces a change in attenuation of about 0·05 in 100 dB so that the error in skin depth must be less than 20% to keep the uncertainty in attenuation equal to the assumed uncertainty in bore diameter.

The second problem is also mechanical, i.e. to measure displacement of the piston without error due to backlash or lost movement. Relative changes in attenuation are often needed to be known within 0·001 dB, which for an attenuation rate of 2·5 dB/mm implies an uncertainty in piston position less than 0·4 μm. Recent advances in technology have presented an elegant solution to this problem as a small retrore-flector can be mounted directly on the piston carriage and a laser interferometer used to give a fringe count with digital readout. A third problem is associated with the frequency sensitivity of a microwave piston attenuator. As shown in Section 4.2.1, the frequency sensitivity is about $9·31 \times 10^{-11}$ dB/mm/Hz, i.e. for an attenuation value of 100 dB caused by a traverse of 40 mm, the sensitivity is $3·72 \times 10^{-9}$ dB/Hz. During the period 1940–1950, the frequency of the source was liable to drift by a few MHz and was usually known only to a similar accuracy by the use of cavity wavemeters. A frequency un-certainty of 2·7 MHz would cause an attenuation uncertainty of 0·01 dB. Fortunately, modern techniques of stabilising microwave sources and measuring their frequency by frequency counters can reduce this error by several orders of magnitude.

Fourthly, the fact that the waveguide beyond cut-off presents an almost purely reactive impedance leads to a mismatch problem at both input and output. Attempts to minimise the effect by using lossy cable or resistive discs were only partly successful and, moreover, entailed the penalty of increased minimum insertion loss and therefore more powerful sources. The incorporation of ferrite isolators, either side of the waveguide beyond cut-off, enables a good match to be presented to generator and load with only a small penalty in increased insertion loss.

Consideration of these factors indicated to C.R. Ditchfield that a 9 GHz waveguide beyond cut-off attenuator was feasible as an independent standard of attenuation, which could be used to compare against other methods such as the modulated subcarrier system. He also decided that a solution to the uncertainty in the value of skin depth would be to incorporate a frequency quadrupler so that the attenuator barrel could be used 'in situ' as a resonator, using the accurate mechanical movement and laser interferometer to accurately establish the resonance positions. It was predicted that this would establish the diameter to better accuracy than conventional metrology. To do this, the Q factor of the cavity would be determined to establish directly the skin depth, thus removing the uncertainty introduced by a lack of precise knowledge of the surface conductivity.

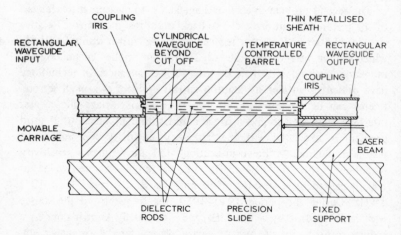

Fig. 4.18 Diagram of RSRE standard attenuator (9 GHz)

The physical form of the attenuator consisted of two standard X-band rectangular waveguides feeding into cylindrical dielectric-filled waveguides. The input waveguide was connected to a thick-walled tube which formed the attenuator barrel (Fig. 4.18). Temperature-controlled water flowed through tubes in the thick walls thus stabilising the temperature of the barrel to about $0.01°C$. The output waveguide was connected to a thin-walled tube, formed by depositing a metallic film on the cylindrical dielectric rod. This tube could slide within the barrel with a minimal clearance of the order of 0.02 mm. Thus, a relative movement of input waveguide with respect to the output waveguide provided the required change in length of the waveguide beyond cut-off. By an appropriate choice of dielectric material for the rod, and diameter of the bore, it was possible to arrange that the only mode which could propagate in the dielectric-filled guide was the H_{11} and that this mode had an attenuation rate of 2·5 dB/mm in the waveguide beyond cut-off. Matching tongues on the dielectric rods protruded through the end walls of the rectangular guides to provide a good impedance match from one propagating guide to the other. Ferrite isolators were used in the rectangular waveguides to prevent undesirable changes in impedance. The mechanical designer decided it would be preferable to move the source in its electrically-screened enclosure rather than incorporate a sliding waveguide junction to compensate for the piston movement.

A detailed analysis was made of the expected sources of error and their magnitudes. Using the barrel as a resonator at 36 GHz, it was calculated that the 'electrical diameter' at this frequency could be determined with an uncertainty of $3·5 \times 10^{-5}$ mm in air, assuming the humidity and temperature of the air were accurately known. As the relevant volume of air was confined in the barrel and substantially, though not completely, isolated from the air in the laboratory, this assumption was somewhat doubtful. By evacuation of the barrel to only a rough vacuum, the uncertainty could be reduced to 2×10^{-5} mm, i.e. 0·02 μm. By measuring the Q values of the resonances the skin depth at 36 GHz could be measured to about 3%. As the skin depth at 9 GHz is twice that at 36 GHz, it was known to a similar percentage uncertainty, i.e. to within 0·04 μm. Thus the electrical diameter at 9 GHz could be estimated to about 10 times better accuracy than by using conventional metrology and estimating the skin depth from the conductivity value. Errors due to the uncertainty in frequency of the microwave source and the velocity of propagation of microwaves were small compared with uncertainties associated with the assessment of the barrel diameter. Assuming that the barrel could be temperature controlled to about 0·01°C, the overall accuracy of attenuation measurement was estimated to be of the order of 0·005 dB in 100 dB.

An instrument based on this design was constructed at RSRE and described at the 1974 conference on precision electromagnetic measurements.[31] The diameter of the bore was chosen to be 10·7 mm and was obtained by gun-drilling a brass cylinder about 50 mm in diameter and 120 mm long. The bore was round to better than 0·25 μm, straight to better than 0·75 μm and had a surface finish of better than 0·05 μm as measured by conventional metrology. With the two dielectric rods in close proximity, the output power was 25 mW indicating only a few dB minimum insertion loss. The attenuation law was linear within

0·01 dB at about 17 dB, and within 0·001 dB at 20 dB, which is a considerable improvement over the earlier designs. A preliminary comparison with other methods of measuring attenuation at 9 GHz indicated agreement within a few thousandths of a decibel over a range of 25 dB. This comparison was made difficult by instabilities in the superheterodyne receiver used, and, at the time of writing, a much more stable receiver is being developed. This will enable a more precise evaluation to be made of the extent of agreement between the two methods.

4.3 References

1 MONTGOMERY, C.G.: 'Technique of microwave measurements', Vol. 11 *in* Radiation Laboratory Series, (McGraw Hill, 1947)

2 CAPPUCCI, J.D.: 'Zero-loss wideband coaxial line variable attenuators', *Microwave J.* 1960, **3**, pp. 49–53

3 MOHR, R.J.: 'New coaxial variable attenuators', *Microwave J.*, 1965, **8**, pp. 99–102

4 SOUTHWORTH, G.C.: 'Principles and applications of waveguide transmission' (Van Nostrand, 1950), pp. 374–376

5 HAND, B.P.: 'Broadband rotary waveguide attenuator', *Electronics,* 1954, **27**, pp. 184–185

6 HAND, B.P.: 'A precision waveguide attenuator which obeys a mathematical law', Hewlett-Packard Journal, 1955, **6**, pp. 1–2

7 MARINER, P.F.: 'An absolute microwave attenuator', *Proc. IEE,* 1962, **109B**, pp. 415–419

8 JAMES, A.V.: 'A high-accuracy microwave attenuation standard for use in primary calibration laboratories', *IRE Trans.*, 1962, **I-11**, pp. 285–290

9 LARSON, W.: 'Table of attenuation error as a function of vane-angle error for rotary vane attenuators'. NBS Technical Note 177, May 1963

10 LARSON, W.: 'Analysis of rotation errors of a waveguide rotary-vane attenuator', *IEEE Trans.*, 1963, **IM-12**, pp. 50–55

11 LARSON, W.: 'Table of attenuation as a function of vane angle for rotary-vane attenuators $(A = -40 \log_{10} \cos \theta)$'. NBS Technical Note 229, January 1965

12 LARSON, W.: 'Gearing errors as related to alignment techniques of the rotary-vane attenuator', *IEEE Trans.*, 1965, **IM-14**, pp. 117–123

13 HOLM, J.D. *et al.*: 'Reflections from rotary-vane precision attenuators', *IEEE Trans.*, 1967, **MTT-15**, pp. 123–124

14 LARSON, W.: 'Analysis of rotationally misaligned stators in the rotary-vane attenuator', *IEEE Trans.*, 1967, **IM-16**, pp. 225–231

15 LOHOAR, J.R.: 'Waveguide rotary attenuators type 6052 series', *Marconi Instrum.* 1968, **11**, pp. 2–5

16 RAHMAN, M.H. and GUNN, M.W.: 'Wave reflections from rotary-vane attenuators', *IEEE Trans.*, 1969, **MTT-17**, pp. 402–403

17 LITTLE, W.E., LARSON, W. and KINDER, B.J.: 'Rotary vane attenuator with an optical readout', *J. Res. NBS*, 1971, **75C**, pp. 1–5

18 OTOSHI, T.Y. and STELZRIED, C.T.: 'A precision compact rotary vane attenuator', *IEEE Trans.*, 1971, **MTT-19**, pp. 843–854

19 'High precision rotary vane attenuator', *NBS Tech. News Bull.*, 1972, **56**, p. 169

20 FOOTE, W.J. and HUNTER, R.D.: 'Improved grearing for rotary-vane attenuators', *Rev. Sci. Instrum.*, 1972, **43**, pp. 1042–1043

21 WARNER, F.L., WATTON, D.O. and HERMAN, P.: 'A very accurate X-band rotary attenuator with an absolute digital angular measuring system', *IEEE Trans.*, 1972, **IM-21**, pp. 446–450

22 LARSON, W.: 'The rotary-vane attenuator as an interlaboratory standard', NBS Monograph 144, Nov. 1975

23 'The microwave engineers' handbook' (Horizon House, 1966), p. 67

24 RUSSELL, A.: 'An absolute digital measuring system using an optical grating and shaft encoder', *Instrum. Rev.* 1966, **13**, pp. 234–237 and 283–286

25 GRIESHEIMER, R.H.: 'Technique of microwave measurements' (McGraw-Hill 1947), pp. 679–719

26 HUXLEY, L.G.: 'Waveguides' (Cambridge University Press, 1947), pp. 58–61

27 GORDON-SMITH, A.C.: 'Calibrated piston attenuator for millimetre waves', *Wireless Engineer*, 1949, **26**, pp. 322–324

28 MUNIER, J.: 'Attenuateur Etalon pour christaux detecteurs UHF', *J. Phys. (France)* 1955, **16**, pp. 429–430

29 GIORDANO, A.B.: 'Design analysis of a TM-mode piston attenuator', *Proc. IRE*, 1950, **38**, pp. 545–550

30 BROWN, J.: 'Corrections to the attenuation constants of piston attenuators', *Proc. IEE*, 1949, **96**, Pt. 3, pp. 491–495

31 DITCHFIELD, C.R.: 'An X band signal generator with WBCO attenuator', *in* 'Precision electromagnetic measurements' *IEE Conf. Publ.* **113**, 1974, pp. 7–9

Chapter 5

Power ratio methods for measuring attenuation

Methods of attenuation measurement that rely on the direct determination of power ratios will be described first as they are more closely related to the basic definition of attenuation than the techniques described in Chapters 6 to 12.

5.1 Straight-forward single power meter method

Fig. 5.1 shows a simple system that can be used for measuring attenuation by the power ratio method. This system can be assembled using waveguide components, coaxial components or a mixture of both. It uses a solid state microwave source (see Chapter 10) and a thin-film thermo-electric (TFT) power meter.[1-4] As a microwave amplitude stabiliser is not used, the solid state source should be operated from a highly stabilised power unit and its bias lead should be surrounded by lossy material to eliminate microwave leakage. Good frequency stability (a few parts in 10^6 per day) can be achieved by using an impatt oscillator which is locked to a high Q invar cavity.[5] Thin-film thermo-electric power meters detect signals over a large frequency range, so the microwave bandpass filter is needed to eliminate harmonics present in the output from the microwave source. To minimise the mismatch error discussed in Chapter 2, the matching units are adjusted until an almost perfect match (v.s.w.r. $< 1·005$) is seen looking in either direction from the insertion point. The isolators are used so that the matching conditions are not affected if any changes are made to the source or power meter. A high resolution swept frequency reflectometer of the type described by Hollway and Somlo[6] is a very good instrument to use when carrying out these matching operations. If a reflectometer is not available, a high grade standing wave detector can be used.

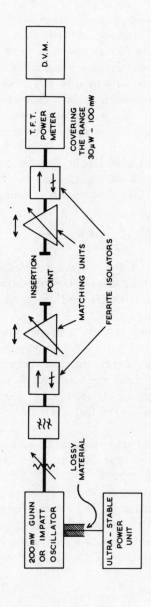

Fig. 5.1 Simple system for attenuation measurement by the power ratio method

Good repeatability is required from the connectors at the insertion point, so it is important to make sure that they are in good mechanical condition, clean and correctly aligned before each mating.

If a TFT power head covering the range $30\,\mu$W to $100\,$mW is used, as shown in Fig. 5.1, maximum dynamic range can be obtained by using a microwave source with an output of about $200\,$mW, as the loss through the filter, the matching units and the two isolators will be in the region of $3\,$dB. If an ultra-sensitive TFT power head covering the range $0\cdot3\,\mu$W to $1\,$mW is used, a $2\,$mW microwave source is entirely adequate and the only disadvantage of using such a combination is that the system becomes 100 times more sensitive to external microwave interference.

Let the digital voltmeter reading be V_1 when the connectors at the insertion point are joined directly together and V_2 when the device under test is inserted. Then the attenuation through the device under test is given by

$$A = 10 \log_{10} \frac{V_1}{V_2} \tag{5.1}$$

With this system, only relative powers are needed so the efficiency of the TFT head need not be known. The lead between the TFT head and the TFT amplifier is flexible and can be moved about without detriment to the accuracy.

Careful measurements made on a simple system of this type at RRE gave results which were in error by only $0\cdot002\,$dB at $3\,$dB and $0\cdot028\,$dB at $20\,$dB. These errors arise firstly because the output e.m.f. of a TFT head is not exactly proportional to the microwave input power, secondly because small errors occur when a power meter is switched from one range to another and thirdly because drifts occur in the equipment.

The TFT power meter can be replaced by a self-balancing thermistor power meter.[7] The latter type drifts more but deviates less from the required square law characteristic.

The effect of changes in the output power of the source will now be investigated. Assume that the TFT power meter is perfectly matched and is an exact square law device with a proportionality constant K_3. Let the output power of the source be P when V_1 is measured, and $P + \Delta P$ when V_2 is measured. Then, the apparent value of the attenuation is

$$A_m = 10 \log_{10} \frac{V_1}{V_2} = 10 \log_{10} \frac{K_3 P}{|s_{21}|^2 K_3 (P + \Delta P)} \tag{5.2}$$

whereas the true value, given by eqn. 2.10 is

$$A_t = 10 \log_{10} \frac{1}{|s_{21}|^2} \tag{5.3}$$

Thus, the error ΔA_p in the attenuation measurement, due to the small change ΔP in the input power, is given by

$$\Delta A_p = A_m - A_t = 10 \log_{10} \frac{1}{\left(1 + \dfrac{\Delta P}{P}\right)}$$

$$\approx -4 \cdot 343 \frac{\Delta P}{P} \tag{5.4}$$

Therefore, an increase of only 0·1% in the output power of the source between the two parts of the measurement gives an attenuation error of $-0·0043$ dB. A 0·1% change in the sensitivity of the TFT power meter would produce the same error.

In a similar manner, it can be shown that the error ΔA_z in an attenuation measurement due to zero drift is given by

$$\Delta A_z \approx -4 \cdot 343 \; \frac{\text{drift in output voltage between the two parts of the measurement}}{\text{output voltage with device under test inserted}} \tag{5.5}$$

Summarising, it can be seen that this very simple system has a dynamic range of about 40 dB, its accuracy is adequate for a great many purposes but it is not good enough for use in a standards laboratory and the results obtained with it are seriously affected by changes in the output power from the source.

The points mentioned in this section about leakage suppression, harmonic filtering, precise matching at the insertion point and connector repeatability apply to all methods of attenuation measurement. Deviation from the required detector law is a common source of error that occurs in many of the measurement techniques described in this book.

Leakage from an attenuation measuring system can be traced rapidly with a small horn that is connected, with a flexible cable or flexible waveguide, to a very sensitive superheterodyne receiver tuned to the source frequency. Leakage can be eliminated by bandaging up any imperfect connectors or components with copper adhesive tape, aluminium foil, flat plastic bags filled with steel wool or any other convenient flexible lossy material.

A panoramic receiver containing a swept YIG filter is a very good instrument for detecting microwave harmonics.

A novel 'match checker' has been described recently by Hollway and Somlo.[8] It can also be used to set up the matching units. The procedure is as follows: a length of precision waveguide is introduced at the insertion point and an attenuator-reflector combination is moved back and forth inside it by a nylon thread. The matching unit adjacent to the reflector is adjusted until the transmitted power stays constant. The attenuator and reflector are then interchanged and the other matching unit is adjusted in the same way.

Reliable information about the repeatability of microwave connector pairs is scarce. However, some values have been given by Bergfried and Fischer[9] and Skilton.[10] Further values are given in various unpublished papers by L.J.T. Hinton of the British Calibration Service. Table 5.1 gives some of the results gleaned from these three sources. The values given in the second column are *upper limits* for the repeatability of clean connector pairs that are in good mechanical condition and are not subjected to any stresses due to misalignment or transverse loads.

The linearity of the power meter can be checked by placing a 3 dB or 6 dB switched coupler (see Chapter 13) at the insertion point and measuring its value at several different settings of the attenuator that appears between the solid state source and the filter in Fig. 5.1. If the power meter is acting as a perfect square law detector, the measured values for the switched coupler will all be the same.

Table 5.1   Repeatability of connector pairs

	Type of connector pair	Repeatability
GR900	14 mm coaxial connector pair	0·003dB up to 8 GHz
APC-7	7 mm coaxial connector pair	0·01dB up to 18 GHz
Type-N	7 mm coaxial connector pair	0·02dB up to 18 GHz
SMA	3·5 mm coaxial connector pair	0·02dB up to 18 GHz
WG15	square flanges, hand lapped and used without dowels	0·005dB at 7·3 GHz
WG15	square flanges machine lapped to a mirror finish and used with precision dowels	0·0005dB at 7·3 GHz
WG15	square flanges used with crushable copper shims and precision dowels	0·0005dB at 7·3 GHz

5.2 Dual channel power ratio method

To overcome the critical dependence on source power discussed in the previous section, Stelzried, Reid and Petty[11, 12] devised the dual channel equipment shown in Fig. 5.2. Two power meters are now used, one before and the other after the insertion point, and a Kelvin Varley

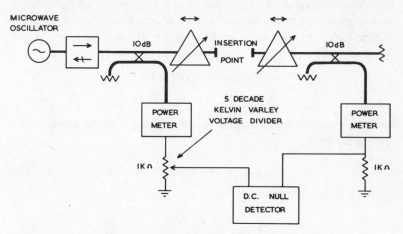

Fig. 5.2 Dual channel d.c. substitution system

voltage divider is used to obtain a null both before and after inserting the device under test.* If both power meters are adjusted to read exactly zero with the microwave oscillator switched off and if they both have perfect square law characteristics, the Kelvin-Varley setting for a null is independent of the output power from the microwave oscillator. This arrangement bears a resemblance to a Wheatstone bridge. With such a bridge, it is well known that the balance point is not affected by changes in the e.m.f. of the battery that drives it.

Let the input to the left-hand power meter be P_{LH}, and let the input to the right-hand power meter be P_d, with the connectors at the insertion point joined directly together, and P_u with the unknown in position. Let the Kelvin-Varley settings for a null without and with the unknown be denoted by D_d and D_u respectively. Then, if K_3 is the proportionality constant for each power meter, the balance equations are

$$K_3 P_{LH} D_d = K_3 P_d \qquad (5.6)$$

and

$$K_3 P_{LH} D_u = K_3 P_u \qquad (5.7)$$

---

*For brevity, the device under test is frequently called the 'unknown' in this book

The attenuation through the unknown is therefore given by

$$A = 10 \log_{10} \frac{P_d}{P_u} = 10 \log_{10} \frac{D_d}{D_u} \qquad (5.8)$$

Great care must be taken to set the zero of the right-hand power meter very accurately. With no microwave input, let us assume that an incorrect zero setting gives an output voltage V_0, from the right-hand power meter. When the same unknown as before is measured, let the Kelvin-Varley settings now be denoted by D'_d and D'_u. Then, the balance equations become:

$$K_3 P_{LH} D'_d = K_3 P_d + V_0 \qquad (5.9)$$

$$K_3 P_{LH} D'_u = K_3 P_u + V_0 \qquad (5.10)$$

and the error in the attenuation measurement due to the wrong zero setting is given by

$$\Delta A_{wz} = 10 \log_{10} \frac{D'_d}{D'_u} - A \qquad (5.11)$$

$$= 10 \log_{10} \frac{1 + \dfrac{V_0}{K_3 P_d}}{1 + \dfrac{V_0}{K_3 P_u}} \qquad (5.12)$$

$$= 10 \log_{10} \frac{1 + \dfrac{V_0}{K_3 P_d}}{1 + \dfrac{V_0}{K_3 P_d} \, \text{antilog} \dfrac{A}{10}} \qquad (5.13)$$

Fig. 5.3 shows how ΔA_{wz} varies with A for three different values of $V_0/K_3 P_d$. To keep the attenuation error below 0·0043 dB up to 20 dB, the zero setting error must be less than 0·001% of the output from the right-hand power meter when the connectors at the insertion point are joined directly together.

In the original version of this equipment made by Stelzried, Reid and Petty, two modified Hewlett Packard type 431 B power meters were used. The short term jitter of this equipment was about 0·0004 dB peak-to-peak and a typical value for its long term drift was 0·0015 dB/h. A precision rotary attenuator with an NBS calibration was measured on this equipment and, up to 16 dB, the results did not differ from the NBS ones by more than 0·1% of the attenuation value in dB. Equipment of this type is now commercially available. Stelzried and Oltmans[13] have carried out precision insertion loss measurements at 90 GHz using this technique. Otoshi *et al*[14] have compared this method with both a reflectometer technique and an a.f. substitution technique. Results obtained by the three methods on lengths of stainless steel waveguide, with a loss of about 0·05 dB, agreed to within 0·0006 dB.

Skilton[15] has made further improvements to this technique by adding a tuned reflectometer with a zero-setting switch to the arrangement shown in Fig. 5.2. A detailed description and analysis of this

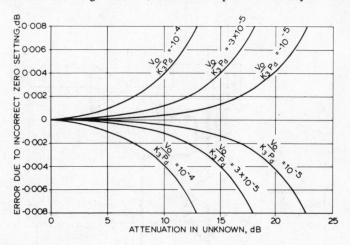

Fig. 5.3 Error due to incorrect zero setting in the right-hand power meter

additional configuration is given in Chapter 9. Skilton[15] has also devised a method of determining and correcting for the non-linearity in the power meters. Five completely independent measurements done by Skilton at 7·3 GHz on a WG15 switch and directional coupler yielded a mean value of 0·1290 dB with 99·7% confidence limits of ± 0·0014 dB and a maximum difference from the mean of 0·0002 dB.

5.3 Systems using amplitude stabilisation of the source

Instead of using two power meters in a bridge arrangement to cancel out power variations arising in the source, it is possible to remove these variations with an amplitude stabiliser. The most stable attenuation measuring system that has been described in the literature up to the present time uses this technique. It was developed by Engen and Beatty[16] at the National Bureau of Standards in 1959. A block diagram of this equipment is shown in Fig. 5.4. The output from the microwave oscillator is stabilised[17] with a servo system containing a self-balancing bolometer bridge in an ultra stable water bath,[18,19] a low drift d.c. amplifier and a ferrite variable attenuator. The power that is trans-

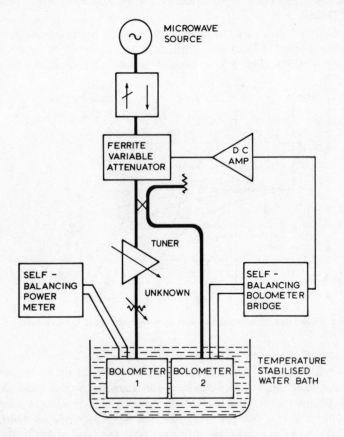

Fig. 5.4 Engen and Beatty's power ratio attenuator calibrator

mitted through the unknown is measured on another self balancing bolometer bridge[20] which also has its vital components immersed in the ultra-stable water bath. The long term stability of this equipment is 0·0002 dB over a 3 hour period and it gives attenuation measurement accuracies on a rotary vane attenuator that rise from 0·0001 dB to 0·06 dB over the range 0·01 dB to 50 dB.

Slightly different measurement techniques are used below and above 3 dB and these will now be described. The changes in the d.c.

power level applied to bolometer 1 during an attenuation measurement are shown in Fig. 5.5.

Let P_z and P_a denote the microwave powers absorbed by bolometer 1 when the rotary attenuator under test is set at its datum position and at an incremental value of A dB, respectively. Also, let W_0, W_z and W_a represent the d.c. bias powers needed in bolometer 1 to balance the bridge when the microwave input powers to it are 0, P_z and P_a, respectively. Then

$$P_z = W_0 - W_z \qquad (5.14)$$

$$P_a = W_0 - W_a \qquad (5.15)$$

and

$$A = 10 \log_{10} \frac{P_z}{P_a} = 10 \log_{10} \frac{W_0 - W_z}{W_0 - W_a} \qquad (5.16)$$

For values of attenuation greater than 3 dB, the power differences $W_0 - W_z$ and $W_0 - W_a$ are measured directly, by backing off the appropriate bias current with a constant current generator.[20]

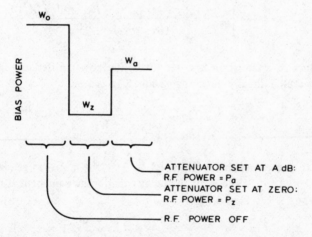

Fig. 5.5 Three bias power levels associated with Engen and Beatty's method

Systematic and random erros in the measurement of the power differences are caused by: inexact balancing of the bridge, deviations from the correct resistance values in the bridge, errors in measuring the bias currents, noise and drift in the amplifier, fluctuations in the backing off current etc. To account for all of the systematic errors.

Let $W_0 - W_z$ be measured as $(W_0 - W_z)(1 + \epsilon_{0z}) \pm P_{0z}$ and let

$W_0 - W_a$ be measured as $(W_0 - W_a)(1 + \epsilon_{0a}) \pm P_{0a}$

Then, the error in determining the attenuation is given by

$$\Delta A_{>3db} = 10 \log_{10} \frac{(W_0 - W_z)(1 + \epsilon_{0z}) \pm P_{0z}}{(W_0 - W_a)(1 + \epsilon_{0a}) \pm P_{0a}} - \log_{10} \frac{(W_0 - W_z)}{(W_0 - W_a)}$$

(5.17)

$$= 10 \log_{10} \frac{1 + \epsilon_{0z} \pm \dfrac{P_{0z}}{(W_0 - W_z)}}{1 + \epsilon_{0a} \pm \dfrac{P_{0a}}{(W_0 - W_a)}}$$

(5.18)

When all of the error terms are very small, it is readily seen that eqn. 5.18 simplifies to

$$\Delta A_{>3db} \approx 4 \cdot 343 \left\{ \epsilon_{0z} - \epsilon_{0a} \pm \frac{P_{0z}}{(W_0 - W_z)} \mp \frac{P_{0a}}{(W_0 - W_a)} \right\}$$ (5.19)

From eqns. 5.16 and 5.19, we get

$$\Delta A_{>3dB} \approx 4 \cdot 343 \left\{ \epsilon_{0z} - \epsilon_{0a} \pm \frac{P_{0z}}{(W_0 - W_z)} \mp \frac{P_{0a} \text{ antilog } (A/10)}{(W_0 - W_z)} \right\}$$

(5.20)

For values of attenuation less than 3 dB, the power difference $W_a - W_z$ is measured directly. Rearranging eqn. 5.16, we get

$$A = 10 \log_{10} \frac{1}{1 - \dfrac{W_a - W_z}{W_0 - W_z}}$$

(5.21)

Owing to the power meter errors, let $W_a - W_z$ be measured as $(W_a - W_z)(1 + \epsilon_{az}) \pm P_{az}$. The error in the attenuation measurement is now seen to be

$$\Delta A_{<3dB} = 10 \log_{10} \frac{1 - \dfrac{W_a - W_z}{W_0 - W_z}}{1 - \dfrac{(W_a - W_z)(1 + \epsilon_{az}) \pm P_{az}}{(W_0 - W_z)(1 + \epsilon_{0z}) \pm P_{0z}}}$$

(5.22)

$$\approx 4 \cdot 343 \frac{\dfrac{W_a - W_z}{W_0 - W_z}}{1 - \dfrac{W_a - W_z}{W_0 - W_z}} \left\{ \epsilon_{az} - \epsilon_{0z} \pm \frac{P_{az}}{(W_a - W_z)} \mp \frac{P_{0z}}{(W_0 - W_z)} \right\}$$ (5.23)

From eqns. 5.21 and 5.23, we finally get

$$\Delta A_{<3\text{dB}} \approx 4\cdot343 \left\{ \left[\text{antilog}\left(\frac{A}{10}\right) - 1 \right] \left[\epsilon_{az} - \epsilon_{0z} \mp \frac{P_{0z}}{(W_0 - W_z)} \right] \right.$$
$$\left. \pm \frac{P_{az} \text{ antilog } (A/10)}{(W_0 - W_z)} \right\} \qquad (5.24)$$

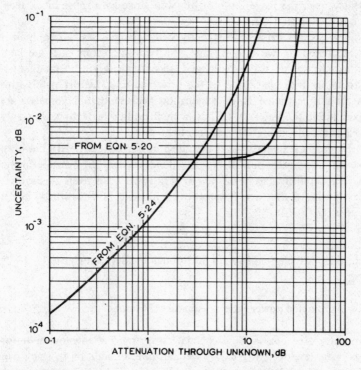

Fig. 5.6 Uncertainties for two different measurement techniques used by Engen and Beatty

In Fig. 5.6, $\Delta A_{<3\text{dB}}$ and $\Delta A_{>3\text{dB}}$ are plotted against A for the following typical values:

$$P_{0z} = P_{0a} = P_{az} = 0\cdot1\,\mu\text{W}$$
$$W_0 - W_z = 10\,\text{mW}$$
$$\epsilon_{0z} - \epsilon_{0a} = \epsilon_{az} - \epsilon_{0z} = 0\cdot001$$

The following features should be noted:

(*a*) The two curves cross over at 3 dB

(*b*) As A approaches zero, $\Delta A_{<3dB}$ approaches a value of $4 \cdot 343\, P_{az}/$ $(W_0 - W_z) = 0 \cdot 000043\, dB$

(*c*) Above 20 dB, $\Delta A_{>3dB}$ rises rapidly due to the last term in eqn. 5.20.

Due to this steep rise, the measurements reported earlier by Engen and Beatty, over the range 20 to 50 dB, were done by a gauge block technique (see Section 8.3.4).

The exceptionally good stability of Engen and Beatty's equipment is attributed to two factors: first, the two bolometers in it are kept in an ultra-stable water bath, and, second, there are no modulation circuits in it to contribute to the jitter and drift. Unfortunately, this method is restricted in its application because both bolometers are fixed inside the water bath. Thus, without additional features, it is only well-suited for calibrating variable attenuators.

Equipment similar to that described by Engen and Beatty has been set up at Physikalisch-Technische Bundesanstalt in West Germany. A very detailed analysis of the errors in it has been given by Bayer.[21] He concludes that its systematic component of uncertainty in dB is given by

$$\Delta A_u \approx 10^{-4}\,(1 + A_u) \qquad (5.25)$$

where A_u is the attenuation in dB through the device under test.

5.4 Automated power ratio attenuator calibrators

An automated coaxial dual channel power ratio attenuator calibrator has been developed at RRE by Hepplewhite. In addition to using two channels to minimise the effect of source power variations, the source is amplitude stabilised with a *p-i-n* diode feedback loop (see Section 10.2.4) and is stabilised in frequency to within ± 1 part in 10^8 per hour. The band 275 MHz to 8 GHz is covered by two powerful lighthouse triode oscillators. Wollaston wire bolometers are used in the two power meters, which are of the self-balancing type. The voltages across the two self-balancing bridges (with the r.f. on and off when the unknown is both in and out) and the frequency are recorded automatically on punched paper tape, using an ultra-linear digital voltmeter, a digital frequency meter and a data logger. The results are worked out on a central computer. By taking a large number of readings rapidly in this way, drift errors are eliminated without using a water bath.

The fixed coaxial components are rigidly clamped to a steel table and the ones that have to be moved about are attached to pneumatic supports[22] so they can be joined together easily without any strain on the connectors. The accuracy of this equipment is similar to that achieved by Engen and Beatty.[16]

An automatic attenuation measuring equipment that can be readily assembled from commercially available items has been described by Edwards.[4] This contains a desk calculator, a programmable signal generator, a two-resistor power splitter,[23] two digital thin film thermoelectric power meters and a printer. The unknown can be measured at hundreds of frequencies in a few minutes to an accuracy of $\pm 0{\cdot}2\,dB$. The dynamic range can be extended to 70 dB by replacing one of the TFT heads with a Schottky diode detector.[24]

5.5 References

1 LUSKOW, A.A.: 'This microwave power meter is no drifter', *Marconi Instrum.* 1971, **13**, pp. 30–34

2 JACKSON, W.H.: 'A thin film/semiconductor thermocouple for microwave power measurements', *Hewlett Packard Journal*, 1974, **26**, pp. 16–18

3 LAMY, J.C.: 'Microelectronics enhances thermocouple power measurements', *Hewlett-Packard Journal*, 1974, **26**, pp. 19–24

4 EDWARDS, A.P.: 'Digital power meter offers improved accuracy, hands-off operation, systems compatibility', *Hewlett Packard Journal*, 1975, **27**, pp.2–7

5 WILSON, K. *et al.*: 'A novel, high stability, high power impatt oscillator', Proceedings of the European Microwave Conference, A6/2:1–A6/2:4, 1971

6 HOLLWAY, D.L. and SOMLO, P.I.: 'A high resolution swept-frequency reflectometer', *IEEE Trans.*, 1969, **MTT-17**, pp. 185–188

7 ADAM, S.F.: 'Microwave theory and applications' (Prentice Hall, 1969), chap. 5

8 HOLLWAY, D.L. and SOMLO, P.I.: 'The match checker – a simple instrument for matching impedances in waveguide and coaxial systems', *IEEE Trans.*, 1974, **MTT-22**, pp. 560–561

9 BERGFRIED, D. and FISCHER, H.: 'Insertion loss repeatability versus life of some coaxial connectors', *IEEE Trans.*, 1970, **IM-19**, pp. 349–353

10 SKILTON, P.J.: 'A technique for determination of loss, reflection and repeatability in waveguide flanged couplings', *IEEE Trans.*, 1974, **IM-23**, pp. 390–394

11 STELZRIED, C.T. and PETTY, S.M.: 'Microwave insertion loss test set', *IEEE Trans.*, 1964, **MTT-12**, pp. 475–477

12 STELZRIED, C.T., REID, M.S. and PETTY, S.M.: 'A precision DC potentiometer microwave insertion loss test set', *IEEE Trans.*, 1966, **IM-15**, pp. 98–104

13 STELZRIED, C.T. and OLTMANS, D.A.: 'Precision insertion loss calibration at 90 GHz', *IEEE Trans.*, 1969, MTT-17, pp. 233–234

14 OTOSHI, T.Y. *et al.*: 'Comparison of waveguide losses calibrated by the DC potentiometer, AC ratio transformer and reflectometer techniques', *IEEE Trans.*, 1970, **MTT-18**, pp. 406–409

15 SKILTON, P.J.: 'A reflectometer and power ratio technique for the measurement of low values of waveguide attenuation', *IEEE Trans.*, 1976, **IM-25**, pp. 307–311

16 ENGEN, G.F. and BEATTY, R.W.: 'Microwave attenuation measurements with accuracies from 0·0001 to 0·06 decibel over a range of 0·01 to 50 decibels, *J. Res. NBS*, 1960, **64C**, pp. 139–145

17 ENGEN, G.F.: 'Amplitude stabilization of a microwave signal source', *IRE Trans.*, *1958*, **MTT-6**, pp. 202–206

18 LARSEN, N.T.: '50 microdegree temperature controller', *Rev. Sci. Instrum.*, 1968, **39**, pp. 1–12

19 HARVEY, M.E.: 'Precision temperature controlled water bath', *Rev. Sci. Instrum.*, 1968, **39**, pp. 13–18

20 ENGEN, G.F.: 'A self-balancing direct current bridge for accurate bolometric power measurements', *J. Res. NBS*, 1957, 59, pp. 101–105

21 BAYER, H.: 'An error analysis for the RF-attenuation measuring equipment of the PTB applying the power method', *Metrologia*, 1975, **11**, pp. 43–51

22 HEPPLEWHITE, L.K. and HARRIS, N.: 'Microwave component support facilitating accurate connector alignment free from strain', *Electron. Lett.*, 1975, **11**, (24), pp. 575–577

23 JOHNSON, R.A.: 'Understanding microwave power splitters', *Microwave J.*, 1975, **18**, pp. 49–51 and 56

24 PRATT, R.E.: 'Very low level microwave power measurements', *Hewlett Packard Journal*, 1975, **27**, pp. 8–10

Chapter 6

R.F. substitution

In the r.f. substitution method, the attenuation through the unknown is measured by comparison with a standard microwave attenuator which is operated at the same frequency as the device under test. The standard microwave attenuator is usually either a very accurate wave-guide rotary-vane attenuator or a precise microwave piston attenuator. Both of these devices have already been described in Chapter 4.

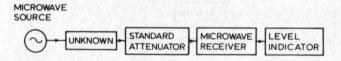

Fig. 6.1 Series r.f. substitution system

For this method of measurement, several different configurations can be used. The simplest is a series connection of the source, the unknown, the standard attenuator, a receiver and a level indicator (see Fig. 6.1). As the attenuation in the unknown is increased, the attenuation in the standard is reduced so that the receiver output stays constant. To achieve high accuracy with this arrangement, it is essential to have both an ultra-stable source and an ultra-stable receiver. This disadvantage can be overcome by placing the unknown and the standard attenuator in parallel with each other and using a null detector (see Fig. 6.2). Magic tees or 3 dB couplers can be used to split and recombine the microwave power. To achieve a balance in a system of this type it is necessary to have a constant loss phase shifter in one of the channels. If a superheterodyne receiver and a powerful source are used with either of these arrangements, a dynamic range in excess of 100 dB can be achieved. Furthermore, the detector law is unimportant because the output is always adjusted to the same value with the series configuration and to a null with the parallel one. This is a most important advantage because many of the attenuation measurement techniques described in this book rely on a particular detector law being maintained throughout the operating range of the instrument, e.g. the i.f. substitution and modulated sub-carrier methods rely on linear mixing, the simple power ratio method shown in Fig. 5.1 is dependent on the output voltage from the thin-film thermo-electric power meter staying directly proportional to the r.f. input power and the simple a.f. substitution systems give erroneous results if the bolometers or diode detectors in them deviate from the required square law characteristic.

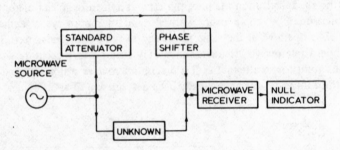

Fig. 6.2 Parallel r.f. substitution system

When a large dynamic range is required from a series r.f. substitution system, two buffered precision rotary attenuators in tandem can be used as the reference standard. The minimum loss through such an arrangement is very small so almost full use can be made of the power available from the source. A microwave piston attenuator, that is operated only over its linear region, has a residual attenuation of about 30 dB, so it is more suitable for parallel substitution than series substitution.

The series configuration has the advantage of not needing a constant loss phase shifter and it is suitable for calibrating the attenuators in signal generators.

A more detailed block diagram of a series r.f. substitution system with a superheterodyne receiver is shown in Fig. 6.3. The output power from the source should be kept as constant as possible and an excellent amplitude stabiliser for doing this has been described by Engen.[1] Matched isolators are required on both sides of the standard attenuator and the unknown to minimise mismatch errors. The automatic frequency control (AFC) loop is needed to keep the intermediate

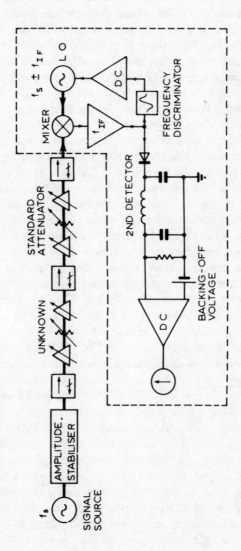

Fig. 6.3 More detailed diagram of a series r.f. substitution attenuator calibrator

frequency constant. Great care must be taken to keep the gain of the i.f. amplifier as constant as possible by using negative feedback (if possible), temperature stabilisation and high class regulation of all voltage supplies. A simple voltmeter connected straight across the output terminals of the second detector has insufficient resolution to enable the output to be set to exactly the same reading each time. It is therefore necessary to back-off the output voltage with a very stable reference voltage and then, after further amplification, display very small changes in the output voltage on a centre-zero meter.

In a system of this type, some form of AFC is essential. If it is not used, frequency drifts in the signal source and the local oscillator will cause the i.f. signal to drift up and down the sides of the i.f. amplifier response curve and serious errors will occur in the attenuation results. However, the AFC loop shown in Fig. 6.3 is only one of many possible solutions to this problem. If measurements are only required at one frequency, the two conventional microwave oscillators shown in Fig. 6.3 could be replaced by two quite separate crystal-controlled transistor varactor sources whose output frequencies differ by f_{IF}. With this solution, great care must be taken to ensure that no errors are caused by spurious signals in the outputs from the transistor varactor sources. A much more versatile solution is the use of two microwave synchronisers to impart crystal stability to each of the microwave oscillators shown in Fig. 6.3. Microwave synchronisers are commercially available from a few firms. A typical one will give a frequency stability of 1 part in 10^8 per hour to any voltage-controlled oscillator operating in the range 1 to 40 GHz. A unit of this type phase locks the microwave oscillator, to which it is connected, to a very high harmonic of a tunable crystal oscillator. Again, great care must be taken to make sure that no errors are caused by unwanted harmonics from the comb generators in the synchronisers.

A simple way of getting round this AFC problem is to use the equipment shown in Fig. 6.4 in place of the equipment enclosed by the dotted area in Fig. 6.3. This arrangement is described in great detail in Sucher and Fox.[2] A sawtooth waveform with a repetition frequency of about 500 Hz is used to sweep the local oscillator over a range of about 10 MHz centred on the frequency $f_s + f_{IF}$. Then, near the middle of each sweep, the correct intermediate frequency is produced and a pip emerges from the video amplifier. The output from this video amplifier is applied to the Y-plates of an oscilloscope and the sawtooth waveform is applied to the X-plates. Each time the attenuator undergoing calibration is changed, the standard attenuator is altered until the pip on the oscilloscope screen has the same amplitude as before. The resetting accuracy can be improved considerably by increasing the video amplifier gain and adjusting the Y shift voltage until only the top part of the pip appears on the oscilloscope screen. With this sawtooth modulation system, frequency drifts in the microwave oscillators cause the pip to move about horizontally, but this is unimportant and does not produce errors in the attenuation measurements.

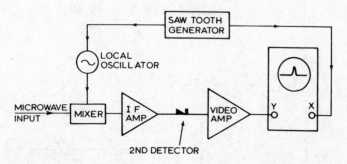

Fig. 6.4 Swept frequency receiving technique that eliminates the need for an AFC loop

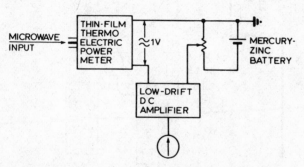

Fig. 6.5 Simple backing-off circuit that enables small changes to be seen in the output from a power meter

A spectrum analyser contains all of the equipment shown in Fig. 6.4 and, if such an instrument is available, it can be used as the receiver for r.f. substitution measurements.

If extreme dynamic range is not required, the superheterodyne receiver shown in Fig. 6.3 can be replaced by a microwave power meter or a crystal video receiver. Due to its very low drift, a thin film thermoelectric power meter is particularly suitable for this role and a simple backing-off arrangement that can be used with it is shown in Fig. 6.5.

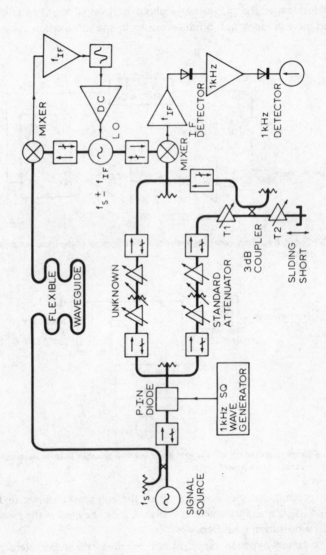

Fig. 6.6 Detailed diagram of a parallel RF substitution attenuator calibrator

The type of AFC system shown in Fig. 6.3 is not suitable for use with a parallel r.f. substitution system, because the local oscillator comes off lock whenever a null is approached. One way of getting round this difficulty is shown in Fig. 6.6. A strong signal is now always fed into the AFC mixer and the local oscillator frequency is not affected at all by adjustments to the microwave bridge.

It is quite a problem to find a satisfactory constant loss phase shifter for use in a parallel r.f. substitution system. A rotary waveguide phase shifter is quite unsuitable as the attenuation through such a device varies by about 0·4 dB as it is revolved. About the best solution to this problem is the arrangement shown in Fig. 6.6. The phase shifter is formed with a 3 dB directional coupler, a precision sliding short circuit and two tuners. The theory of this configuration is given in Chapter 9. Tuner T2 is adjusted to make the coupler directivity infinite, and then tuner T1 is adjusted until the sliding short circuit looks into a perfect match. Movement of the sliding short circuit now provides the necessary phase shifts with an attenuation variation of less than 0·001 dB. In Fig. 6.6, the null detection system is made extra sensitive by square wave modulating the microwave input to the bridge and following the second detector with a 1 kHz tuned amplifier.

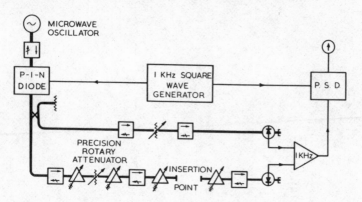

Fig. 6.7 Dual channel series r.f. substitution system

Fig. 6.7 shows a dual channel series r.f. substitution system which was described recently by Larson and Campbell.[3,4] It is not affected by source power variations and it does not require a constant loss phase shifter so it overcomes the disadvantages of both the series and parallel configurations. It should be noted that the subtraction of the two signals occurs at 1 kHz instead of at the microwave frequency. As the two microwave diodes in it are used as detectors and not mixers, the dynamic range is only about half of that provided by the systems shown in Figs. 6.3 and 6.6. The precision rotary attenuator which is used as the reference standard in this equipment has already been described[5] in Chapter 4. The drift of this equipment is typically 0·0002 dB over a period of 12 minutes.

References

1   ENGEN, G.F.: 'Amplitude stabilisation of a microwave signal source', *IRE Trans.*, 1958, **MTT-6**, pp. 202–206
2   SUCHER, M. and FOX, J.: 'Handbook of microwave measurements – Vol. 1', (Polytechnic Press, 1963) chap. 7
3   LARSON, W. and CAMPBELL, E.: 'Microwave attenuation measurement system (RF series substitution)'. Digest of IEEE International Convention, pp. 434–435, 1971
4   LARSON, W. and CAMPBELL, E.: 'Microwave attenuation measurement system (series substitution)', NBS Technical Note 647, Feb. 1974
5   LITTLE, W.E., LARSON, W. and KINDER, B.J.: 'Rotary vane attenuator with an optical readout', *J. Res. NBS*, 1971, **75C**, pp. 1–5

Chapter 7

I.F. substitution

7.1 General

The i.f. substitution method of attenuation measurement relies on comparison with a precision i.f. attenuator, which is usually a piston attenuator. This method has been widely used during the last 30 years. It gives good accuracy and a large dynamic range (≈ 100 dB). Two main configurations have been used: series i.f. substitution[1,2] and parallel i.f. substitution.[3-15]

The series configuration is in principle the simpler of the two, but it has certain disadvantages which will be discussed later. A stabilised r.f. or microwave source is connected through the attenuator under test (set at reference zero and required value in turn) to a linear frequency convertor with an i.f. output. This i.f. output is amplified by about 30 dB to make up for the initial loss in the i.f. piston-attenuator which follows. The preamplifier has to be as linear as the frequency convertor if error is to be avoided. Beyond the standard piston attenuator are the main i.f. amplifier, second detector and output level indicator (Fig. 7.1). The output level is maintained at a convenient constant value. Each increase in the unknown attenuation is counteracted by an observed decrease in the standard. Errors and their reduction will be described later.

The parallel configuration, although in principle more complicated, has been the more popular arrangement. It is particularly useful for calibrating attenuators in standard-signal generators used for testing all types of radio receivers, but it can also be used to calibrate 2-port attenuators if a stabilised r.f. or microwave source is provided. Here, this source will be regarded as part of the parallel system (Figs. 7.2 and 7.3). The stabilised microwave source is connected through the attenuator under test (set at reference zero and required value in turn)

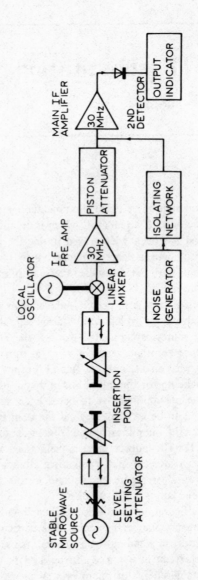

Fig. 7.1 Series i.f. substitution system

to a linear frequency convertor with an i.f. output. This output is compared with the output from the standard piston attenuator which is driven by a stable i.f. source. An i.f. amplifier with a sensitive output

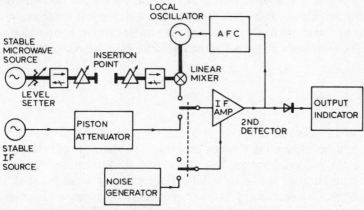

Fig. 7.2 Parallel i.f. substitution system with manual switching

level meter is used as a transfer instrument to indicate equality between i.f. outputs from the frequency convertor and the standard piston attenuator. To use this transfer instrument as a comparator, it is necessary either to switch it between the two i.f. outputs manually (in which case the i.f. amplifier gain must be extremely stable) (Fig. 7.2) or to

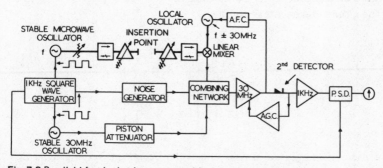

Fig. 7.3 Parallel i.f. substitution system with electronic switching

switch it electronically and rapidly by square-wave modulation of the microwave and i.f. sources, which are thereby switched on and off alternately. The two switched inputs are combined in a hybrid network to prevent interaction, and a synchronous detector is used with a null-indicator (Fig. 7.3). Again, errors and their reduction will be described later.

Typical values of i.f. in use are 30 MHz and 60 MHz, and the methods are used for r.f. and microwave frequencies from about 10 MHz to 40 GHz.

The piston attenuator used in an i.f. substitution system is normally calibrated directly in decibels, so an attenuation change in the unknown can be found by simply subtracting the two piston attenuator readings that correspond with the initial and final settings of the unknown.

7.2 Frequency conversion

In an i.f. substitution system, it is vitally important that there should be a linear relationship between the i.f. output voltage from the mixer and the r.f. signal voltage that is fed into it. This relationship has been investigated theoretically for a diode mixer by Weinschel et al.[5] using Fourier analysis, by Hollway and Kelly,[10] who made use of an earlier paper by Moullin,[16] and by Gardiner and Bannerjee[17] using an iterative computer technique. The two earlier treatments give essentially the same results. These treatments assume that the diode has a linear forward characteristic and infinite reverse resistance. It is also assumed that the i.f. output voltage follows the envelope of the resultant input wave. Let the *peak* values of the signal, local oscillator and i.f. voltages be denoted by V_s, V_0 and V_{if}, respectively. Also, let the d.c. component of the rectified current be represented by I_{dc}. Then, these treatments show that

$$V_{if} \approx K_1 V_s \left\{ 1 - \frac{1}{8} \left(\frac{V_s}{V_0} \right)^2 \right\} \tag{7.1}$$

and

$$I_{dc} \approx K_2 V_0 \left\{ 1 + \frac{1}{4} \left(\frac{V_s}{V_0} \right)^2 \right\} \tag{7.2}$$

where K_1 and K_2 are usually constants. Using eqn. 7.1, a general expression for the error in an attenuation measurement due to a departure from linearity in the mixer can be derived as follows: let an attenuation increase of A dB in the unknown cause the input signal voltage to fall from $V_{s,max}$ to $V_{s,min}$. Then

$$A = 20 \log_{10} \frac{V_{s,max}}{V_{s,min}} \tag{7.3}$$

However, the measured attenuation change will be

$$A_m = 20 \log_{10} \frac{K_1 V_{s,max} \left\{1 - \frac{1}{8}\left(\frac{V_{s,max}}{V_0}\right)^2\right\}}{K_1 V_{s,min} \left\{1 - \frac{1}{8}\left(\frac{V_{s,min}}{V_0}\right)^2\right\}} \qquad (7.4)$$

The error in the attenuation measurement due to mixer non-linearity is

$$\Delta A = A_m - A = 20 \log_{10} \frac{\left\{1 - \frac{1}{8}\left(\frac{V_{s,max}}{V_0}\right)^2\right\}}{\left\{1 - \frac{1}{8}\left(\frac{V_{s,min}}{V_0}\right)^2\right\}} \qquad (7.5)$$

When $V_{s,max}/V_0$ is small, eqn. 7.5 can be simplified to

$$\Delta A \approx -1 \cdot 086 \left(\frac{V_{s,max}}{V_0}\right)^2 \left\{1 - \left(\frac{V_{s,min}}{V_{s,max}}\right)^2\right\} \qquad (7.6)$$

Let γ denote the ratio in dB of the local oscillator voltage to the maximum signal voltage. Then

$$\gamma = 20 \log_{10} \frac{V_0}{V_{s,max}} \qquad (7.7)$$

From eqns. 7.3, 7.6 and 7.7, we finally get

$$\Delta A = -\frac{1 \cdot 086}{\text{antilog}(\gamma/10)} \left\{1 - \frac{1}{\text{antilog}(A/10)}\right\} \qquad (7.8)$$

Fig. 7.4 shows how ΔA varies with γ for different values of A. These curves show that for highly linear operation (error $< 0 \cdot 001$ dB), the maximum signal voltage which is fed into the mixer must be kept at least 30 dB below the local oscillator voltage that is applied to the mixer.

If self bias is used (i.e. bias is obtained by rectification of the local oscillator voltage, using a load resistor) and if the diode characteristic for forward applied voltages is non-linear, then because the bias is increased when the signal V_s is increased to its maximum value, as shown by eqn. 7.2, the factor K_1 in eqn. 7.1 will change when V_s approaches its maximum value and the i.f. non-linearity will be considerably worse than that indicated by eqn. 7.1 with K_1 constant. Experiments with increasing values of bias resistor[12] have confirmed this. Therefore, a self-bias resistor should be avoided and a fixed d.c. bias of low effective source resistance substituted. This arrangement also leads to improved noise performance, (important with low-level signals), and Siddle and Harris[12] found that the non-linearity with large signals is then given by eqn. 7.1 with K_1 constant.

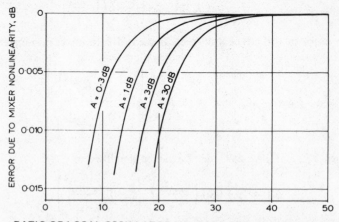

Fig. 7.4 Mixer non-linearity error

A simple mixer circuit such as is used for the lower microwave frequency bands[12] is shown in Fig. 7.5. The lumped-element signal input circuit is symbolic only: a distributed circuit or a cavity may actually be used.

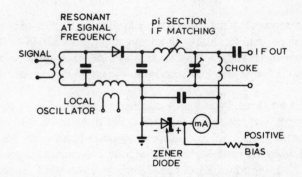

Fig. 7.5 Typical mixer circuit used at lower microwave frequencies

If it is desired to keep ΔA below 0·005 dB and allow V_s to rise to 0·5 V peak, it follows from eqns. 7.6 and 7.7 that $\gamma = 23·4$ dB and

$V_0 = 7.4$ V, when A is large. With a local oscillator voltage of this magnitude, a diode that will stand a reverse potential of at least 15 V is needed. For frequencies up to about 2 GHz, efficient solid-state diodes with this reverse rating are available. Many of the diodes that have been designed for use as microwave mixers will only stand a reverse potential of about 2 V, and, for the same non-linearity of 0·005 dB, we now find that the upper limits on V_0 and V_s are 1 V peak and 68 mV peak, respectively. If a voltage step up ratio of 2 occurs between the mixer signal input terminal and the mixer diode, and a dynamic range of 100 dB is required, the receiver must now be capable of giving an adequate signal to noise ratio with a peak signal voltage at the mixer input terminal of only 0·34 μV.

At microwave frequencies, a balanced mixer, using two diodes, has advantages: it almost cancels out the amplitude modulation noise on the local oscillator signal and it greatly reduces the amount of local oscillator power that is fed back into the signal source.

In any mixer, single or balanced, a resonant circuit at the input, tuned to the signal frequency, is desirable to make the source impedance at the image frequency independent of the impedance of the device under test which may vary during an attenuation calibration.

Eqn. 7.2 shows that the d.c. component of the rectified current starts to increase if V_s becomes large, e.g. if V_s is raised from zero to a value 20 dB down on V_0, I_{dc} will increase by 0·25%. Thus, by observing I_{dc}, the threshold of severe non-linearity can be detected.

There are several approximations associated with the theory given in this section and mixer non-linearity errors worse than those predicted by eqn. 7.8 have been reported from time to time by calibration laboratories. Thus, when using an i.f. substitution system, the mixer linearity should always be checked in the manner described at the end of Section 5.1.

A detailed practical investigation of mixer linearity by Hollway and Kelly[10] gave results that are in good agreement with eqn. 7.8. Yell[18] has carried out linearity measurements at both L and X-bands on Schottky barrier mixer diodes and he found smaller departures from linearity than those reported by Hollway and Kelly.[10] The theoretical results obtained by Gardiner and Banerjee[17] are in close agreement with Yell's experimental results.

There appears to be a need for a further detailed investigation into this subject of mixer non-linearity, with particular attention being paid to the biasing arrangement, single versus balanced mixers and the effects of an input filter tuned to the signal frequency.

7.3 Errors and their reduction in the series method

A source of systematic error peculiar to the series method is associated with the initial loss in the i.f. piston attenuator, which appears in a particularly acute form because the departure from linearity of the attenuator scale cannot be reduced by stabilisation of the launching current as it can be in the parallel method. Unless a special calibration of the non-linear part of the attenuator scale has been made, this initial loss will be in excess of 30 dB and this has to be made up by a linear amplifier preceding the i.f. piston attenuator.

When the device under test is set to its highest attenuation value, the i.f. signal from the mixer will be weak and the piston will be adjusted to give a low value of attenuation. Under these conditions, noise originating in the mixer and preamplifier will be important and there will be a poor signal to noise ratio at the second detector. When the attenuation through the device under test is reduced to a low value, the i.f. signal from the mixer will be much larger, the piston will be set to give a high value of attenuation and the signal to noise ratio at the second detector will then be very high and will be mainly determined by the noise from the main i.f. amplifier. If identical readings are obtained on the output meter in these two cases, there will be an error in the indicated value of attenuation because the two signal levels will actually be different by an amount equal to the effective amplitude of the output noise in the former case. Kaylie[2] has shown that this error can be eliminated by injecting noise into the main amplifier when the piston is providing high values of attenuation. The amplitude of this injected noise is carefully adjusted to give an output signal to noise ratio equal to that obtained when the piston is set at its lowest value and the noise generator is switched off. The noise is injected after the piston attenuator through an isolating network to prevent the noise source loading the input of the main i.f. amplifier and producing a further error when it is switched on and off. By using this noise injection technique, the dynamic range of the series method can be extended to 100 dB if sufficient care is paid to the design of the launching and pick-up coils in the piston attenuator.

Automatic frequency control of the local oscillator is not as important in the series method as it is in the parallel method. However, its provision can save the operator a lot of time.

7.4 Errors and their reduction in the parallel method

The parallel method has a unique source of systematic error associated with the need to switch alternately between the i.f. output of the frequency convertor and the output from the piston attenuator, in order to compare them for equality.

The use of manual i.f. switching demands a high degree of gain stability in the i.f. amplifier, as well as very good amplitude and frequency stabilities in the r.f. and i.f. sources, over a time longer than the switching period. An advantage is that a narrow i.f. bandwidth can be used provided that the local oscillator is furnished with automatic frequency control to keep the i.f. signal at the peak of the response curve of the amplifier. A narrow bandwidth reduces the noise at the detector.

Square-wave modulation can succeed with poorer long-term gain stability than is required for manual switching, but, with a modulation frequency of, say, 1 kHz, the overall bandwidth has to be large enough to enable the signals to be amplified faithfully, and 1 MHz is probably the minimum satisfactory value. This larger bandwidth tends to result in greater noise at the detector than is obtained with the narrower bandwidth possible with manual switching.

With both forms of switching, it is necessary to equalise the noise levels at the detector when switching from the mixer output to the piston attenuator output, because the former output is usually the noisier of the two. This is done by adding noise to the output from the piston attenuator. In the case of square-wave modulation, a combining network is necessary to prevent undesirable interaction between the mixer, piston-attenuator and noise-generator outputs. This combining network usually contains hybrid transformers.[10]

When using square wave modulation, the output from each source should remain constant to within a specified uncertainty (e.g. 0·001 dB) throughout each 'on' period. It is difficult to ensure that this requirement is being met. One possible solution is to use a type of *p-i-n* diode, which remains matched as the bias current through it is varied,[19] and connect it into a circuit which enables it to perform as a fast-acting amplitude stabiliser throughout each 'on' period and as a total absorber throughout each 'off' period.

With square wave modulation, yet another problem can arise. Large spikes can be generated at the change-over points of the square wave and these spikes sometimes have different magnitudes at the positive and negative edges. They can give an error because, in a phase sensitive detector, the mean values of the input signals on alternate

half-cycles are subtracted from each other. This source of trouble
can be eliminated by supressing the receiver throughout each change-
over period.

7.5 Comments on various present-day i.f. substitution systems

A large number of i.f. substitution attenuator calibrators are now being
used throughout the world, and Table 7.1 gives the overall accuracies
that are provided by two of the popular commercially available equipments.

Table 7.1   Comparison of accuracies of i.f. substitution systems

Attenuation step	Overall accuracy of commercially available series i.f. substitution system	Overall accuracy of commercially available parallel i.f. substitution system
dB	dB	dB
10	± 0·04	± 0·01
20	± 0·05	± 0·02
40	± 0·07	± 0·04
80	± 0·20	± 0·27
100	± 0·40	± 0·50

Both of these equipments use noise injection, automatic frequency
control and phase sensitive detection.

Hollway and Somlo[13] have recently added several refinements to
the parallel i.f. substitution system that is used in the Australian National Standards Laboratory. These refinements include: a switched
reference path to reduce drift errors, a flexible waveguide in the measuring path to accommodate attenuators of different lengths, a steady
flow of dry air through the system to remove moisture, and automatic
control of the i.f. oscillator power output for fine balancing of the two
channels.

An all-solid-state parallel i.f. substitution system has been developed
by Yell[14] at the National Physical Laboratory in England, for use as
a national standard. The piston attenuator in it[20] is mounted vertically.
The piston is supported on air bearings to prevent contact with the
cylinder wall. The piston displacements are measured with a laser

interferometer. This piston attenuator is of the type shown in Fig. 3.29*b*. A 60 MHz crystal-controlled oscillator is attached to the piston. The launching and pick-up coils are on quartz formers and multi-strip mode filters are used. The cylinder is made of electroformed copper, which was deposited at a rate of 0·025 mm/h on to a stainless steel mandrel with a nominal diameter of 50·5854 mm and with circularity and straightness accurate to within 0·25 μm. The piston can be moved over a distance of 200 mm by a recirculating-ball leadscrew that is driven by a stepper motor. Fine adjustments down to 0·08 μm are achieved with a leadscrew driven wedge, which is moved by a second stepper motor. The conductivity of the electroformed copper was determined from Q measurements in a manner similar to that described in Section 3.3.4. This piston attenuator has a range of 120 dB and a resolution of 0·0002 dB. Its accuracy has been stated to be 0·001 dB in 120 dB.

When Yell's equipment is used at X-band, the signal and local oscillator sources are varactor-tuned Gunn oscillators and these are phase locked one harmonic apart to the 60 MHz crystal-controlled oscillator. These Gunn oscillators are operated from extremely stable power supplies and have amplitude stabilities of better than 0·001% per minute. Planar Schottky-barrier diodes are used in a balanced mixer. Switching between the mixer and piston attenuator outputs is done at 500 Hz by an array of series and shunt *p-i-n* diodes. The 60 MHz i.f. amplifier contains multiple differential-pair integrated circuits. Input and interstage tuned circuits give an i.f. bandwidth of 1·5 MHz.

Automatic gain control (a.g.c.) is used with a time constant of 1 s in the control line. The mean output voltage from the second detector is kept constant by this a.g.c. system, but, owing to the 1 s time constant, the 500 Hz amplitude modulation on the i.f. signal is not diminished. The i.f. response curve is almost independent of the a.g.c. voltage and the switching transients are gated out. The second detector is followed by a 500 Hz tuned amplifier and a phase-sensitive detector. The stability of the receiver is of the order of 0·0001 dB per 10 minutes.

A fully automatic parallel i.f. substitution system, covering the range 1–18 GHz, has been described by Weinert and Weinschel.[15] It has a digital readout with a resolution of 0·001 dB and a single-step dynamic range of 110 dB. This equipment does not contain a piston attenuator. Its place has been taken by a box of extremely low temperature coefficient π type attenuators, which are switched in and out automatically by very repeatable mercury-wetted reed relays. These π type attenuators are not standards in their own right. Each one is carefully calibrated against a very precise piston attenuator that is fitted with a laser interferometer. The errors are stored in a programmable read-only semiconductor memory that is incorporated in the receiver. These errors are then automatically subtracted in real time from the readout. The test and local oscillator signals are provided by totally independent multi-band, all-solid-state, programmable, frequency stabilised sources. A barretter is used in the signal levelling loop to minimise harmonic errors. The i.f. amplifier is centred on 30 MHz. The signal and 30 MHz sources are switched on and off alternately by 20 kHz square waves. The switched attenuators cover the range 0–140 dB in 1 dB steps, and the three least-significant digits for the readout are obtained from a linear detector and an analogue to digital converter. With automatic balancing, the repeatability is better than ± 0·01 dB at 60 dB. The absolute accuracy of the reference standard is ± 0·003 dB/10 dB ± 0·001 dB.

7.6 References

1 GRANTHAM, R.E. and FREEMAN, J.J.: 'A standard of attenuation for microwave measurements', *Trans. AIEE*, 1948, **67**, pp. 329–335

2 KAYLIE, H.L.: 'A new technique for accurate RF attenuation measurements', *IEEE Trans.*, 1966, **IM-15**, pp. 325–332

3 GAINSBOROUGH, G.F.: 'A method of calibrating standard-signal generators and radio frequency attenuators', *J. IEE*, 1947, **94**, Pt. 3, pp. 203–210

4 HEDRICH, A.L., *et al.*: 'Calibration of signal generator output voltage in the range 100 to 1000 megacycles', *IRE Trans.*, 1958, **I-7**, pp. 275–279

5 WEINSCHEL, B.O., *et al.*: 'Relative voltmeter for VHF/UHF signal generator attenuator calibration', *IRE Trans.*, 1959, **I-8**, pp. 22–31

6 WEINSCHEL, B.O.: 'An accurate attenuation measuring system with great dynamic range', *Microwave J.*, 1961, **4**, pp. 77–83

7 TURNER, R.J.: 'Equipment for the measurement of the insertion loss of waveguide networks in the frequency range 3·8–4·2 Gc/s', *Proc. IEE*, 1962, **109B**, Suppl. 23, pp. 775–782

8 WALLIKER, D.A.J.: 'Resistive attenuators and their calibration', *Proc. IEE*, 1962, **109B**, Suppl. 23, pp. 791–795

9 CLARK, R.F., and DEAN, B.J.: 'A precision attenuation standard for X-band' *IRE Trans.*, 1962, **I-11**, pp. 291–293

10 HOLLWAY, D.L. and KELLY, F.P.: 'A standard attenuator and the precise measurement of attenuation', *IEEE Trans.*, 1964, **IM-13**, pp. 33–44

11 RUSSELL, D.H.: 'An unmodulated twin-channel microwave attenuation measurement system', *ISA Trans.*, 1965, **4**, pp. 162–169

12 SIDDLE, R.W.A. and HARRIS, I.A.: 'A CW comparator for precision RF attenuators', *Radio & Electron. Eng.*, 1968, **35**, pp. 175–181

13 HOLLWAY, D.L. and SOMLO, P.I.: 'The reduction of errors in a precise microwave attenuator calibration system', *IEEE Trans.*, 1973, **IM-22**, pp. 268–270

14 YELL, R.W.: 'The design of signal sources and receiver for use in a precise microwave attenuation calibrator', *IEEE Trans.*, 1974, **IM-23**, pp. 371–374

15 WEINERT, F.K. and WEINSCHEL, B.O.: 'A 1–18 GHz attenuator calibrator', *CPEM Digest*, 1976, pp. 94–96

16 MOULLIN, E.B.: 'The detection by a straight line rectifier of modulated and heterodyne signals', *Wireless Eng. & Exp. Wireless*, 1932, **9**, pp. 378–383

17 GARDINER, J.G. and BANERJEE, A.R.: 'Conversion-loss stability and gain compression in Schottky-barrier mixers', *Electron. Lett.* 1970, **6**, (26), pp. 829–830

18 YELL, R.W.: 'Signal-compression performance of Schottky-barrier diodes in microwave mixers', *Electron. Lett.*, 1969, **5**, (16), pp. 360–361

19 HUNTON, J.K. and RYALS, A.G.: 'Microwave variable attenuators and modulators using PIN diodes', *IRE Trans.*, 1962, **MTT-10**, pp. 262–273

20 YELL, R.W.: 'Development of a high precision waveguide beyond cut-off attenuator', *CPEM Digest*, 1972, pp. 108–110

Chapter 8

A.F. substitution

In this chapter, attenuation measuring equipment which uses bolometers as square law detectors is discussed first. A description is then given of a precise attenuator calibrator which has been developed in Canada by Clark. Finally, a detailed treatment of modulated subcarrier attenuator calibrators is presented.

8.1 A.F. substitution systems using bolometers as square law detectors

The simplest a.f. substitution system is shown in Fig. 8.1. The microwave oscillator is square wave modulated at 100 Hz and the bolometer is used as a square law detector. Its output is passed through a precision audio attenuator and a low noise amplifier and it is then rectified and displayed on a meter. With the two connectors at the insertion point joined directly together, the audio attenuator is adjusted until a convenient reading is obtained on the output meter. The 'unknown' is then inserted and the audio attenuator is readjusted until the same output as before is obtained. Owing to the square law nature of the detector, the microwave attenuation in decibels through the 'unknown' is equal to half the increment on the audio attenuator in decibels. A system of this type has been described by Korewick[1,2] and dual channel versions of it, which are unaffected by amplitude variations in the microwave source, have been described by Sucher and Fox,[3] Finnie et al.[4] and Pakay.[5] Dual channel equipment of this type is commercially available and a typical one has a dynamic range of 25 dB and an accuracy of ± 0·005 dB over the range 0—1 dB. The accuracy falls off to ± 0·1 dB over the range 20—25 dB.

The dynamic range of equipment of this type is determined by detector and amplifier noise at the lower end and by deviations from the

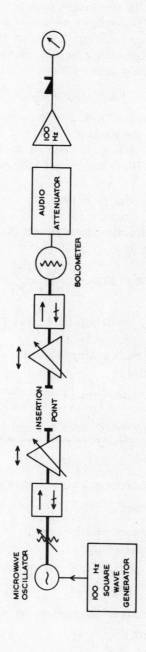

Fig. 8.1 Korewick's series a.f. substitution system

required square law characteristic at the upper end.[6-8] Noise has received a vast amount of attention in other books so it will not be discussed further here. However, we will now consider the effect of a departure from the required square law on an attenuation measurement. Let the bolometer output voltage at the modulation frequency, due to an r.f. input power of p, be given by

$$V_m = Tp + Up^2 \qquad (8.1)$$

where T and U are constants. With a perfect square law characteristic, U would be zero. With the attenuator under test set at zero, let $V_m = V_0$ and $p = p_0$. When it is set to given an attenuation of A dB, let $V_m = V_A$ and $p = p_A$. Then, the measured value for the attenuation is

$$A_m = 10 \log_{10} \frac{V_0}{V_A} \qquad (8.2)$$

and the error in the attenuation measurement cuased by the deviation from square law is given by

$$\Delta A_{sl} = A_m - A = 10 \log_{10} \frac{Tp_0 + Up_0^2}{Tp_A + Up_A^2} - 10 \log_{10} \frac{p_0}{p_A}$$

$$= 10 \log_{10} \left\{ \left(1 + \frac{U}{T} p_0\right) \left(1 + \frac{U}{T} p_A\right)^{-1} \right\} \qquad (8.3)$$

Since $(U/T)p_0$ will be much smaller than unity, and $p_A < p_0$, this last equation can be simplified to

$$\Delta A_{sl} \approx 4 \cdot 343 \frac{U}{T} p_0 \left\{ 1 - \frac{1}{\text{antilog}\,(A/10)} \right\} \qquad (8.4)$$

The resistance of a bolometer, which is biased by a constant current I_B, is given by

$$R = R_0 + J(p + I_B^2 R)^n \qquad (8.5)$$

where $R_0 =$ the cold resistance of the bolometer

$J =$ a constant

and $n \approx 0.9$ for many commercially available bolometers. Using eqn. 8.5, it is shown in Appendix 5 that

$$\frac{U}{T} \approx -\frac{(1-n)}{2T_{B'}^2} \left(\frac{R}{R_0}\right)^2 \approx -\frac{(1-n)}{2P_{dc}} \left(\frac{R}{R_0}\right)^2 \qquad (8.6)$$

where P_{dc} is the d.c. power dissipated in the bolometer. From eqns. 8.4 and 8.6 we get $2I_B^2 R$

$$\Delta A_{sl} \approx -2 \cdot 172 \, (1-n) \frac{p_0}{P_{dc}} \left(\frac{R}{R_0}\right)^2 \left\{1 - \frac{1}{\text{antilog} \, (A/10)}\right\} \quad (8.7)$$

Using the typical values: $n = 0 \cdot 9$, $R_0 = 140 \, \Omega$, and $R = 200 \, \Omega$, Fig. 8.2 shows curves of ΔA_{sl} against A for various values of p_0/P_{dc}. It can be seen that p_0 must be kept 20 dB below P_{dc} to keep the error in the attenuation measurement below 0·005 dB.

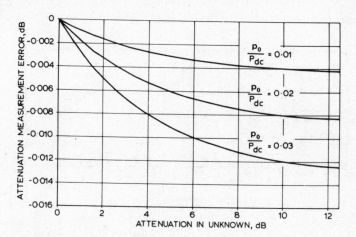

Fig. 8.2 Attenuation measurement error due to deviation from square law in the bolometer

In a system of the type described by Korewick, the bolometer can be replaced by a microwave diode detector that is operated in its square law region. A detailed analysis of deviation from square law in a diode detector has been given by Sorger and Weinschel.[8] They conclude that the bolometer is the better detector for use in this type of equipment.

8.2 A.F. substitution system used at NRC

The attenuation measuring system described in this section has been developed to a very high degree by Clark[9] at NRC, Ottawa. It is similar to an i.f. substitution system, but it uses a 10 kHz inductive voltage divider as the reference standard. A simplified block diagram of this system is shown in Fig. 8.3. The output from the synthesiser is split and fed into harmonic generators and single sideband modulators which produce two microwave signals differing in frequency by 10 kHz. One of these signals is amplitude stabilised, passed through the unknown and then fed into a linear mixer. The other one is used as a local oscillator signal. Thus, a 10 kHz intermediate frequency is produced. At each setting of the unknown, the IVD is adjusted until the rectified output is equal to the voltage of the mercury battery, as indicated by a null reading on the recorder. The changes of attenuation in the unknown are then readily deduced from the settings of the IVD. With this equipment, the short term jitter is about 0·0003 dB peak-to-peak and the estimated systematic error varies from 0·0005 dB at 3 dB to 0·01 dB at 80 dB.

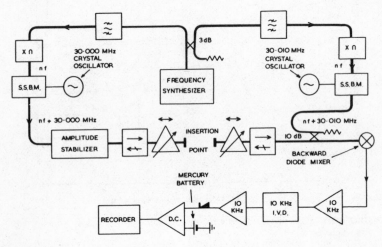

Fig. 8.3 Clark's series a.f. substitution system

8.3 Modulated sub-carrier attenuator calibrators

8.3.1 Introduction
The modulated sub-carrier method of attenuation measurement was devised by Schafer and Bowman[10] in 1962 and it has been analysed further by Little,[11] Oseiko et al.[12] and Kalinski.[13] A simplified block diagram of a modulated sub-carrier system is shown in Fig. 8.4. Basically, it is just a 1 kHz bridge. The 1 kHz sinusoidal oscillator shown on the left is the bridge source. In the left-hand side of the bridge, the 1 kHz signal is stepped down with great precision by two IVDs, which are connected in tandem. In the right-hand side of the bridge, the 1 kHz signal is amplitude modulated onto a microwave carrier, passed through the microwave attenuator under test and then extracted again by a linear microwave balanced mixer. The local oscillator signal reaches the balanced mixer through the waveguide shown on the

right-hand side. In a homodyne system of this type, it is shown in the next section that the amplitude of the i.f. signal emerging from the balanced mixer is proportional to cos θ, where θ is the phase difference between the unmodulated wave and the carrier component of the amplitude modulated wave. Immediately before each attenuation measurement, the rotary phase shifter is adjusted until θ is zero, to maximise the amplitude of the i.f. signal. At each setting of the attenuator under test, the IVDs are adjusted until the bridge is balanced and the changes in attenuation are then deduced from the IVD settings. Any phase changes that occur in the attenuator under test can be read directly from the rotary phase shifter.

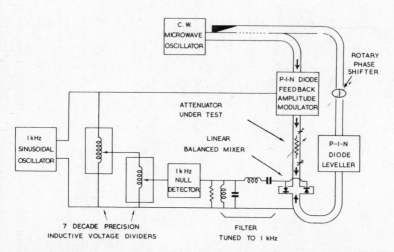

Fig. 8.4 Simplified block diagram of a modulated sub-carrier attenuator calibrator

8.3.2 Detailed description of the RRE modulated sub-carrier systems

Four modulated sub-carrier systems in waveguide sizes 11A, 16, 18 and 22 have been developed at the Royal Radar Establishment for use as British national standards and a detailed block diagram of the X-band one is shown in Fig. 8.5. To obtain the largest possible dynamic range without causing attenuator damage, a 1 W CW two-cavity klystron is used as the microwave source and its output is split into two channels with a 10 dB coupler. The signal passing through the sub-carrier channel is amplitude modulated sinusoidally at 1 kHz to a depth of 80% with a d.c. coupled *p-i-n* diode envelope feedback system which stabilises both the mean output power and the modulation depth. The 1 kHz input to this modulator is obtained from a Baxandall current switching oscillator[14] which is followed by a maximally flat filter. The amplitude

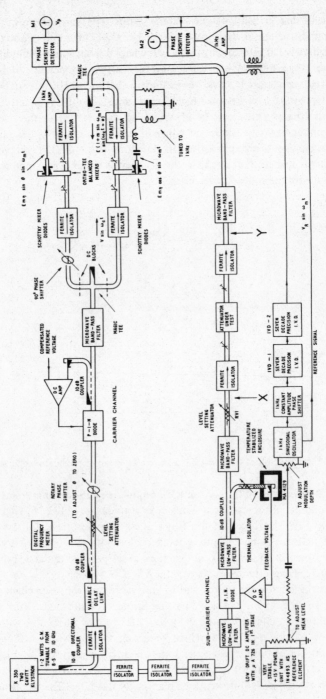

Fig. 8.5 Block diagram of the RRE modulated sub-carrier system used for the precise measurement of attenuation at X-band

stability of this oscillator is better than ± 0·003% per hour and the total harmonic distortion in its filtered output is 0·002%. The d.c. reference voltage that is fed into this envelope feedback modulator has a stability of better than ± 0·001% per hour. Since microwave detector diodes have a temperature coefficient in the region of −0·02 dB/°C, the MA 4129 diode in this modulator has its temperature stabilised to ± 0·01°C This diode has been carefully selected for low flicker noise. Three ferrite isolators are connected in cascade between the first 10 dB coupler and the *p-i-n* diode modulator. These reduce the 1 kHz modulation in the carrier channels to a value well below the noise level. The amplitude modulated signal is passed through a level setting attenuator and the attenuator under test, which is sandwiched between carefully matched isolators. It is then divided between two balanced mixers containing silicon Schottky barrier diodes. The unmodulated signal in the carrier channel is fed, through a variable delay line, a rotary phase shifter and a *p-i-n* diode leveller, into a magic tee where it is similarly split into two parts, one of which is fed directly into the lower balanced mixer while the other is fed, through a 90° phase shifter, into the upper balanced mixer. The lower balanced mixer in Fig. 8.5 corresponds to the balanced mixer shown in Fig. 8.4.

The two signals which are applied to the lower balanced mixer can be written as $E(1 + m \sin \omega_m t) \sin (\omega_c t + \theta)$ and $V \sin \omega_c t$. Combining these two signals vectorially, the amplitude of the resultant is found to be

$$V_A = \{V^2 + E^2(1 + m \sin \omega_m t)^2 + 2VE(1 + m \sin \omega_m t) \cos \theta\}^{1/2}$$

$$= V\{1 + r^2 + 2r \cos \theta\}^{1/2} \tag{8.8}$$

where

$$r = \frac{E}{V}(1 + m \sin \omega_m t) \tag{8.9}$$

In practice, r is deliberately made much smaller than unity, so, neglecting r^2 in comparison with unity and then expanding by the binomial theorem and neglecting all terms except the first two, we get

$$V_A \approx V(1 + r \cos \theta) \approx V + E \cos \theta + Em \cos \theta \sin \omega_m t \tag{8.10}$$

Thus, after linear detection with a conversion factor η, the amplitude of the 1 kHz component in the output from the lower balanced mixer is seen to be $Em\eta \cos \theta$.

The two signals applied to the upper balanced mixer can be written as $E(1 + m \sin \omega_m t) \sin (\omega_c t + \theta)$ and $V \sin (\omega_c t + \psi)$ where ψ is the phase change introduced by the 90° phase shifter. The amplitude of the resultant in this case is

$$V_p \approx \{V^2 + E^2(1 + m \sin \omega_m t)^2 + 2VE(1 + m \sin \omega_m t)\cos(\psi - \theta)\}^{1/2}$$

$$(8.11)$$

Proceeding now in the same way as before, we get

$$V_p \approx V\{1 + r \cos(\psi - \theta)\} \qquad (8.12)$$

When $\psi = \pi/2$, we have

$$V_p \approx V(1 + r \sin \theta) \approx V + E \sin \theta + Em \sin \theta \sin \omega_m t \qquad (813.)$$

so the amplitude of the 1 kHz component emerging from the upper balanced mixer is seen to be $Em\eta \sin \theta$. A phase-sensitive null detector is connected to the output of the upper balanced mixer, and, before each attenuation measurement, the rotary phase shifter is adjusted until there is a zero reading on meter M1. This ensures that θ is always kept at zero and the signal emerging from the lower balanced mixer is therefore always maximised. This arrangement makes the adjustment of θ simpler and quicker because $\cos \theta$ varies slowly with θ in the vicinity of its maximum value. The null detection system which drives meter M2 in Fig. 8.5 corresponds to the null detector shown in Fig. 8.4.

The input stage of the 1 kHz amplifier in the amplitude channel is shown in Fig. 8.6. The amplifier itself consists of a *n-p-n, p-n-p, n-p-n* feedback triple with an a.c. gain of 100 and a d.c. gain of 2. It is operated from an internal mercury-zinc battery. This completely eliminates circulating earth currents and hum pick-up troubles. The push-pull connection of the two mixer diodes enables them to operate with zero bias under all drive conditions. The crystal current can be checked when desired by inserting a 0–1 mA meter into the jack which is normally shorted. The input filter, which consists of a series tuned circuit resonant at 1 kHz followed by a parallel tuned circuit also resonant at 1 kHz, drastically reduces any harmonics present in the mixer output. When a null is achieved, no current flows through the input transformer so there is no loading on the output of the inductive voltage divider. Thus, an essential condition for maximum accuracy from this divider is fully satisfied whenever readings are taken from it. The neutralising circuit formed by the four components between the upper end of the input transformer and the emitter of the BC109 is carefully adjusted so that there is no output at all from the amplifier when large signals of exactly the same amplitude and phase are applied to both ends of the input transformer. This neutralising circuit is necessary to cancel out a small signal which reaches the base of the BC109 under these conditions through stray capacitances in the input transformer.

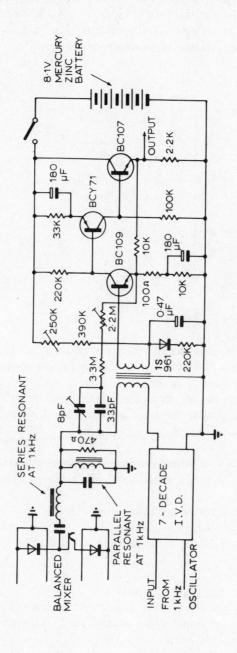

Fig. 8.6 Circuit diagram of the first stage of the 1 kHz amplifier

The later stages of this amplifier contain three more feedback triples, a bandpass filter and two switched gain controls. The amplifier in the phase channel has a single-ended input circuit, but, in all other respects, it is identical with the one in the amplitude channel.

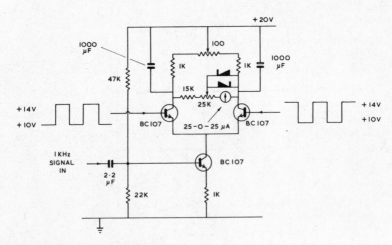

Fig. 8.7 Phase sensitive detector

Many of the phase sensitive detector circuits in use today are very complicated but the two used in this modulated sub-carrier system are extemely simple (see Fig. 8.7). Never-the-less, they have proved to be entirely adequate. The output from the Baxandall oscillator is squared up to provide the switching waveforms for these phase sensitive detectors.

8.3.3 Errors in a modulated sub-carrier attenuator calibrator
The sources of error in a modulated sub-carrier attenuator calibrator are:
(a) imperfections in the inductive voltage dividers
(b) non-linearity in the mixer diodes
(c) imperfect matching of the source and load seen by the attenuator under test
(d) leakage
(e) microwave and 1 kHz harmonics
(f) frequency and power variations in the microwave source
(g) frequency and amplitude variations in the 1 kHz source
(h) variations in the modulation depth
(i) frequency pulling of the microwave oscillator by the rotary phase shifter

(*j*) attenuation variations in the rotary phase shifter
(*k*) circulating earth currents
(*l*) noise
(*m*) variations in the ambient temperature
(*n*) drifts in the electronic circuits
(*o*) incorrect setting of the angle θ.

The steps that have been taken to minimise all of these errors in the equipment shown in Fig. 8.5 will now be described.

Errors in the reference standard have been reduced to a low level by using some of the best inductive voltage dividers that are commercially available and using two in tandem when measuring high values of attenuation. The errors that arise in an IVD are fully analysed in Chapter 3. At low values of attenuation this error is below 10^{-5} dB with a single IVD, and, at 100 dB, it is about 10^{-3} dB with two IVDs in tandem.

Mixer non-linearity in a modulated sub-carrier system has been investigated by Little.[11] At RRE, the mixer non-linearity error is kept below 0·001 dB by setting attenuator RV1 so that the amplitude modulated signal arriving at the balanced mixer is always at least 35 dB and preferably 40 dB below the unmodulated signal that is fed into it. Built-in waveguide switches that are not shown in Fig. 8.5 enable these power levels to be checked easily at any time using a thin film thermoelectric power meter. The mixer linearity is checked by placing a 10 dB switched-coupler (see Chapter 13) at the insertion point and measuring its value at several different settings of RV1. In the truly linear region, the measured values for the switched coupler are almost identical and a satisfactory setting for RV1 can be readily determined.

The mismatch error is fully discussed in Chapter 2. The matching units adjacent to the unknown are adjusted until a v.s.w.r. of less than 1·005 is seen looking in either direction from the insertion point. These matching units and their associated isolators are removed from the modulated sub-carrier system and connected to a high resolution swept-frequency reflectometer[15] for this operation. The isolators used here have reverse attenuations exceeding 40 dB, so it is adequate to place matched loads behind them when the matching is done. If the reflection coefficient magnitudes of the unknown are determined, the upper limit of the mismatch error can be found from the equations or curves in Chapter 2. The mismatch error is often the largest source of error.

Leakage has been reduced to a negligible level by bandaging up all imperfect microwave connectors with aluminium foil, metallised adhesive tape or plastic bags filled with steel wool. Leakage checks are carried out frequently with a small horn which is connected by a flexible coaxial cable to a sensitive superheterodyne receiver. Numerous filters are used to eliminate the microwave and 1 kHz harmonics.

Frequency fluctuations in the microwave source are made unimportant by adjusting the microwave variable delay line until the two halves of the waveguide system form a balanced time bridge. To carry out this adjustment, the source is frequency modulated over a range of several MHz and the outputs from the two 1 kHz amplifiers are applied to the X and Y-plates of an oscilloscope. This gives a propeller-shaped display which reduces to a straight line when the delay is correct. In addition to this precaution, the frequency can be kept within 20 kHz of the required X-band frequency by observing the digital frequency meter and making occasional electronic tuning adjustments to the microwave oscillator.

Power variations in the microwave source are removed by the two p-i-n diode stabilisers. The one in the carrier channel is made insensitive to ambient temperature changes by using a reference voltage that has the same temperature coefficient as the detector diode in it. This reference voltage is obtained from a zener diode and $4 p$-n junction diodes which are mounted in a copper block attached to the detector diode holder.

Being a 1 kHz bridge system, amplitude variations in the 1 kHz source have no serious effect, and frequency variations in this source are not important either because the 1 kHz amplifiers have flat-topped bandpass response curves and phase sensitive detectors are used. Furthermore, both amplifiers are only used as null detectors.

The 1 kHz modulation is introduced by means of the p-i-n diode envelope feedback system which keeps the modulation depth constant.

When the rotary phase shifter is adjusted, its v.s.w.r. varies, but it is prevented from pulling the frequency of the klystron by the ferrite isolator which is placed in front of it. The attenuation through this rotary phase shifter varies by about 0·4 dB as it is revolved, but this variation is removed by the p-i-n diode leveller in the carrier channel.

Circulating earth currents have been eliminated by the appropriate use of isolating transformers and d.c. blocks across the waveguide.

Noise has been minimised by using type 5082–2750 low flicker noise diodes in the mixers, and type BC109 transistors, operated at very low currents[16,17] in the input stages of the 1 kHz amplifiers. In Appendix 6, it is shown that the error in an attenuation measurement due to noise is given by

$$\Delta A_n = 4\cdot 343 \sqrt{\frac{FkT_0}{P_0\tau}\left\{\frac{1}{m^2}+\frac{1}{2}\right\}}\,\text{antilog}\,\frac{A}{20} \qquad (8.14)$$

Where F = noise factor (expressed as a power ratio)
       k = Boltzmann's constant
       T_0 = room temperature in $^{\circ}$K
       P_0 = total power reaching mixer through sub-carrier channel
              when A = 0 dB
       τ = output time constant
       m = modulation depth
       A = attenuation through the unknown in dB.

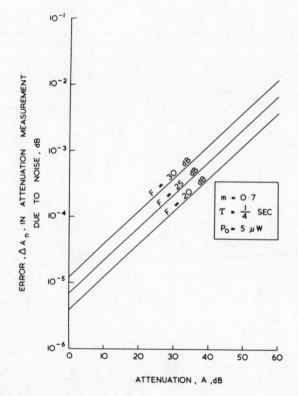

Fig. 8.8 Error due to noise

Fig. 8.8 shows how ΔA_n varies with A for several different values of
F. It can be seen that the error due to noise starts to become important when the attenuation through the unknown reaches about 50 dB.

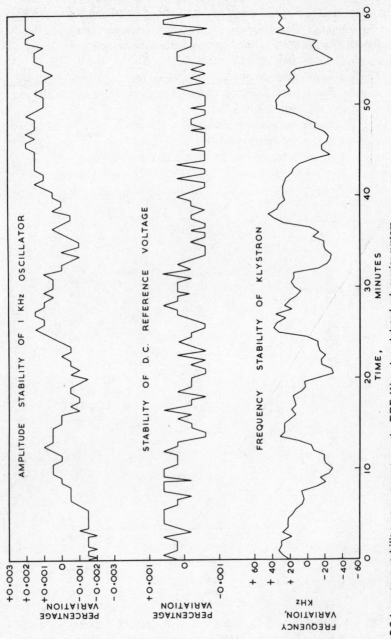

Fig. 8.9 Long term stability measurements on the RRE X-band modulated sub-carrier system

A special technique has been developed at RRE to extend the measurement range to > 100 dB, and this is described in Section 8.3.4.

To minimise the effect of ambient temperature variations, this equipment is installed in a laboratory where the temperature is controlled to ± 0·5°C.

Drifts in the electronic circuits have been kept as low as possible by using low temperature coefficient components, highly stabilised h.t. voltages and negative feedback wherever appropriate. Five long-term stability curves for this equipment are shown in Figs. 8.9 and 8.10. The klystron frequency varies periodically over a range of ± 30 kHz in sympathy with the cyclic variations of ± 0·5°C produced by the laboratory's temperature controller. Over the 30 minute period shown, the drifts at the outputs of the phase and amplitude channels are seen to be ± 0·05° and ± 0·0006 dB, respectively.

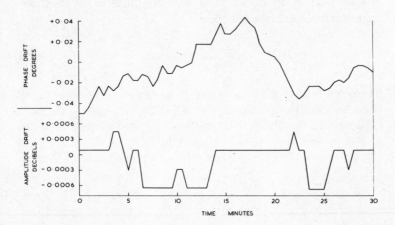

Fig. 8.10 Curves showing the long term stability of the RRE *X*-band modulated sub-carrier system

There is a rather subtle error connected with the setting of the rotary phase shifter. This is associated with the approximations that are made in the theory in Section 8.3.2.

Expanding eqn. 8.11 without any approximations, we get

$$V_p^2 = V^2 + E^2 \left(1 + \frac{m^2}{2}\right) + 2VE \cos (\psi - \theta)$$

$$+ 2VEm \left\{\frac{E}{V} + \cos (\psi - \theta)\right\} \sin \omega_m t - \frac{E^2 m^2}{2} \cos 2\omega_m t \quad (8.15)$$

Thus, the 1 kHz component in the output from the upper balanced mixer, is zero, not when $\psi = 90°$, but when

$$\frac{E}{V} + \cos(\psi - \theta) = 0 \qquad (8.16)$$

During the initial setting-up procedure, with the attenuator under test set at zero, let the amplitude of the modulated signal be denoted by E_0, let θ be set to zero by maximising the output from the lower balanced mixer and let ψ be set so that there is a zero reading on meter M1. Then

$$\frac{E_0}{V} + \cos\psi = 0 \qquad (8.17)$$

and the ratio of the unmodulated wave amplitude to the modulated wave amplitude, expressed in dB, is

$$\gamma = 20\log_{10}\frac{V}{E_0} \qquad (8.18)$$

Suppose now that the unknown is set to give an attenuation of A dB (with no phase change) and let this reduce the amplitude of the modulated wave to E_A. Owing to the relationship given by eqn. 8.16, a small deflection will now appear on meter M1 and let the rotary phase changer be altered by $\Delta\theta$ to return this meter reading to zero. Then

$$\frac{E_A}{V} + \cos(\psi - \Delta\theta) = 0 \qquad (8.19)$$

The output from the lower balanced mixer will now be proportional to $E_A \cos(\Delta\theta)$ and the measured value of the attenuation in dB is given by

$$A_m = 20\log_{10}\frac{E_0}{E_A \cos(\Delta\theta)} \qquad (8.20)$$

However the true value of the attenuation is

$$A = 20\log_{10}\frac{E_0}{E_A} \qquad (8.21)$$

Thus, the error in the attenuation measurement caused by the use of the quadrature channel to set the rotary phase changer, is found from eqns. 8.17 to 8.21 to be

$$\Delta A_\theta = A_m - A = 20 \log_{10}$$

$$\cfrac{1}{\cos\left\{\cos^{-1}\left(-\cfrac{1}{\text{antilog}\,\cfrac{\gamma}{20}}\right) - \cos^{-1}\left(-\cfrac{1}{\text{antilog}\,\cfrac{A+\gamma}{20}}\right)\right\}} \quad (8.22)$$

Fig. 8.11 shows how ΔA_θ varies with γ for different values of A. It is clearly necessary to keep the maximum amplitude modulated signal 35 to 40 dB below the unmodulated signal to make this error negligible. The angle ψ is then close to $90°$.

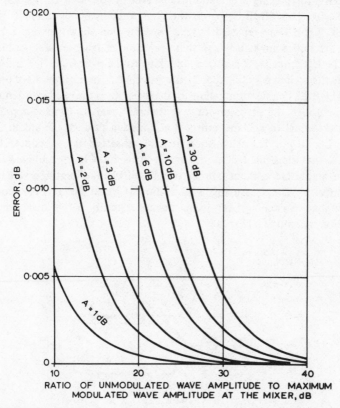

Fig. 8.11 Error caused by using the quadrature channel to set the rotary phase shifter

8.3.4 Special techniques for measuring high values of attenuation with a modulated sub-carrier system

Modulated sub-carrier systems are almost ideal for calibrating variable attenuators. Under normal measurement conditions, the level setting attenuator RV1, shown in Fig. 8.5, is adjusted to about 50 dB to enable highly linear mixing to be obtained. If one wishes to calibrate a variable attenuator up to 80 dB, the range from 0 to 40 dB can be covered in the normal manner. The range from 40 to 80 dB can then be calibrated with 40 dB removed from the level setting attenuator. By using this procedure, there is a high signal to noise ratio throughout the entire calibration.

A special technique has been developed at RRE for making attenuation measurements on short lengths of waveguide.[18] This is fully described in Chapter 9. Another special technique has been devised at RRE for making attenuation measurements on components with a loss exceeding 60 dB.[18] To do this, the components between X and Y in Fig. 8.5 are replaced by the configuration shown in Fig. 8.12. The six tuners are adjusted so that very good matches are seen looking to the left from M, O and Q, and looking to the right from N, P and R, with the switches in the appropriate positions. The various A's shown in Fig. 8.12 denote attenuations in decibels. A_0 is the residual attenuation through the precision rotary attenuator, and A_ϕ is made approximately equal to half the expected attenuation through the unknown, i.e. $A_\phi \approx A_u/2$. The attenuation in the level setter RV1 is reduced by A_ϕ so that the signal fed into the mixer through the lower route has the same amplitude as the maximum signal used when calibrating a variable attenuator. The measurement procedure is based on a gauge block technique. Switching over from the upper route to the lower route gives an attenuation change

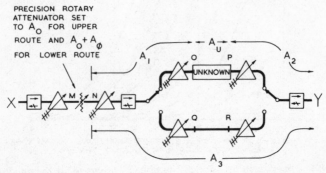

Fig. 8.12 Components inserted between X and Y in Fig. 8.5 when measuring high values of attenuation.

$$A_{m1} = (A_0 + A_1 + A_u + A_2) - (A_0 + A_\phi + A_3) \qquad (8.23)$$

which is measured in the usual way. The unknown is then replaced by a piece of waveguide of the same length with an attenuation A_W.

With the precision rotary attenuator now left fixed at $A_0 + A_\phi$, switching over as before gives an attenuation change

$$A_{m2} = (A_0 + A_\phi + A_1 + A_W + A_2) - (A_0 + A_\phi + A_3) \quad (8.24)$$

From eqns. 8.23 and 8.24 we get

$$A_u = A_{m1} - A_{m2} + A_\phi + A_W \qquad (8.25)$$

A_ϕ is then measured in the usual way, and A_W is determined by the reflectometer technique described in Chapter 9, so A_u can be found. Since A_0, A_1, A_2, A_3 and A_W are low values of attenuation, it follows that $A_{m1} \approx A_u/2$ and A_{m2} is very low.

The variable attenuator shown in Fig. 8.12 should be capable of being reset very accurately. The RRE precision rotary attenuator with an opto-electronic angular measuring system (see Chapter 4) has proved to be excellent for this role.

Attenuations exceeding 100 dB have easily been measured by this technique, and 50 measurements done in this way on an 80 dB coupler had a standard deviation of only 0·0064 dB.

8.3.5 Automatic calibration of rotary vane attenuators on a modulated sub-carrier system

Several automated rotary vane attenuators have been developed at RSRE for use in self-balancing radiometer systems, and it is necessary to calibrate them at all preferred frequencies and repeat the calibrations every few months. Manual calibrations of this type are tedious, they absorb skilled workers for very long periods and the results can be afflicted with human errors, so a simple fully-automatic method of calibrating them has been devised at RSRE and evaluated at X-band.[19] This system will now be briefly described.

The automated rotary vane attenuators were derived from the RSRE rotary vane attenuation transfer standards that are described in Section 13.5. They have been equipped with a servo drive unit, a Moore Reed absolute digital angular measuring system[20] with a resolution of 0·0001° and a 3-stage opto-electronic homing system, which enables the rotor to be driven with great accuracy to three preset angles that can be chosen at will.

Seven changes were made to the X-band modulated sub-carrier system shown in Fig. 8.5 to make it automatic, and its new block diagram is shown in Fig. 8.13. The items in the lower right-hand part of this diagram are shown in greater detail in Fig. 8.14. All of the relay contacts are shown in the unenergised positions. The seven changes were as follows:

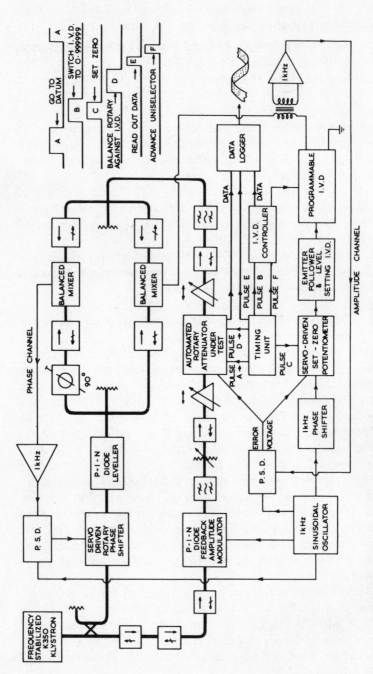

Fig. 8.13 Block diagram of the RSRE X-band automated modulated sub-carrier system

(*a*) the main inductive voltage divider has been replaced by a programmable one with 6 decades (Gertsch Model PRT-D-60200)

(*b*) a frequency stabiliser has been connected to the microwave source. This keeps the frequency stable to within ± 1 part in 10^8 per hour

(*c*) a servo-motor has been fitted to the rotary phase shifter. This is driven by the output from the upper PSD (See Fig. 8.13) and it keeps the phase channel continuously nulled

(*d*) a servo-driven set zero potentiometer has been added

(*e*) a very simple electronic timing unit has been incorporated in it. This produces the six waveforms shown in Fig. 8.13. The durations of these waveforms can be set to any required values by adjusting calibrated variable resistors

(*f*) an IVD controller has been added. This contains 12 rows of decimal to BCD thumbwheel switches, with 6 switches per row. These are set in advance by the operator to the voltage ratios at which calibrations are needed. The BCD signals from these switches are connected, through a 24-level 12-way plug-in uniselector, to 24 transistors that drive the relays in the IVD

(*g*) a data logger has been added.

Further details about the 3-stage opto-electronic homing system are shown in Fig. 8.14. The homing is accomplished with the aid of semicircular discs that are mounted on the same shafts as the encoders. These discs can interrupt infrared beams passing from GaAs sources to photo transistors.

High-speed reed relays, driven by double comparators with hysteresis successively connect the coarse, medium and fine sensors to the servo amplifier (see Fig. 8.14). Relay G is energised until the rotor is within about $\pm 0.3°$ of the desired angle, and relay H is energised until it is within about $\pm 0.01°$ of the desired angle. The homing errors indicated by the digital angular measuring system do not exceed $\pm 0.015°$ with one stage in operation, $\pm 0.001°$ with 2 stages in operation, and $\pm 0.0001°$ with all 3 stages in use.

The automatic procedure will now be described. During period A, contact A/1 is changed over and the rotary attenuator is driven to its datum position by the 3-stage homing system. During period B, contacts B/1 to B/24 are changed over and the 24 transistors are now connected to voltages which set the IVD to 0.999999 (a setting of 1.000000 is not possible with the type of IVD used). While the IVD is at this setting, pulse C changes over contact C/1 and the set-zero potentiometer is now driven in the apppropriate direction until the output from the amplitude channel is zero. During period D, the rotary attenuator is driven until the output from the lower balanced mixer is

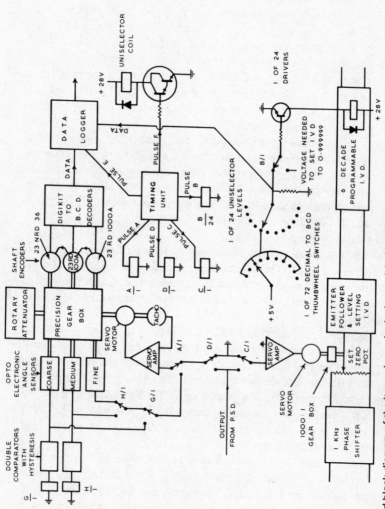

Fig. 8.14 More detailed block diagram of the items shown in the lower right-hand part of Fig. 8.13

equal to the output from the programmable IVD (which depends on the appropriate thumbwheel switch settings). Pulse E energises the data logger and the BCD signals from the digital angular measuring system and the uniselector are now punched on to a paper tape which therefore contains the rotor angles corresponding to a series of voltage ratios. This tape is subsequently processed on a central computer. Pulse F causes the uniselector to advance to its next position and the whole process is then repeated over and over again.

Only the simplest mode of operation has been described. However, several variations can be readily achieved, e.g. each measurement can be repeated any number of times up to 9 by feeding pulse F into a division circuit.

The results achieved without, and with, automation are in good agreement. With automatic operation, the random component of the uncertainty up to 20 dB is typically ± 0·0002 dB (for 9 measurements and a confidence level of 99%).

8.4 References

1 KOREWICK, J.: 'Audio modulation substitution system for microwave attenuation measurements', *IRE Trans.*, 1953, **MTT-1**, pp. 14–21

2 KOREWICK, J.: 'A-M system measures microwave attenuation', *Electronics*, 1954, **27**, pp. 175–177

3 SUCHER, M., and FOX, J.: 'Handbook of microwave measurements – Vol. 1' (Polytechnic Press, 1963), pp. 412–413

4 FINNIE, C.J., *et al.*: 'AC ratio transformer technique for precision insertion loss measurements'. Proceedings of the 19th Annual ISA Conference, preprint 12.6–3–64, Oct. 1964

5 PAKAY, P.: 'A simple method for the accurate measurement of small attenuations', *Period. Polytech.*, 1974, **18**, pp. 105–116

6 SORGER, G.U. and WEINSCHEL, B.O.: 'Precise insertion loss measurements using imperfect square law detectors and accuracy limitations due to noise', *IRE Trans.*, 1955, **I-4**, pp. 55–68

7 WEINSCHEL, B.O.: 'Insertion loss test sets using square law detectors', *IRE Trans.*, 1955, **I-4**, pp. 160–164

8 SORGER, G.U. and WEINSCHEL, B.O.: 'Comparison of deviations from square law for RF crystal diodes and barretters', *IRE Trans.*, 1959, **I-8**, pp. 103–111

9 CLARK, R.F.: 'Superheterodyne measurement of microwave attenuation at a 10 kHz intermediate frequency', *IEEE Trans.*, 1969, **IM-18**, pp. 225–231

10 SCHAFER, G.E. and BOWMAN, R.R.: 'A modulated sub-carrier technique of measuring microwave attenuation', *Proc. IEE*, 1962, **109B**, Suppl. 23, pp. 783–786

11 LITTLE, W.E.: 'Further analysis of the modulated sub-carrier technique of attenuation measurement', *IEEE Trans.*, 1964, **IM-13**, pp. 71–76

12 OSEIKO, A.A., RABINOVICH, B.E. and STOYAKINA, O.V.: 'Measuring attenuation by means of a modulated sub-carrier', *Meas. Tech. (USA)*, 1970, 3, pp. 394–396

13 KALINSKI, J.: 'Further possibilities of the modulated sub-carrier technique for microwave attenuation measurements in industrial applications', *IEEE Trans.*, 1972, IM-21, pp. 291–293

14 BAXANDALL, P.J.: 'Transistor sine-wave LC oscillators', *Proc. IEE*, 1959, 106B, Suppl. 16, pp. 748–758

15 HOLLWAY, D.L. and SOMLO, P.I.: 'A high resolution swept-frequency reflectometer', *IEEE Trans.*, 1969, MTT-17, pp. 185–188

16 FAULKNER, E.A.: 'The design of low-noise audio-frequency amplifiers', *Radio & Electron. Eng.*, 1968, 36, pp. 17–30

17 FAULKNER, E.A. and HARDING, D.W.: 'Some measurements on low-noise transistors for audio-frequency applications', *Radio & Electron. Eng.*, 1968, 36, pp. 31–33

18 WARNER, F.L., HERMAN, P. and JEFFS, T.: 'Special techniques for measuring low and high values of attenuation with a modulated sub-carrier system', *IEEE Trans.*, 1974, IM-23, pp. 381–386

19 WARNER, F.L., WATTON, D.O., HERMAN, P. and CUMMINGS, P.: 'Automatic calibration of rotary vane attenuators on a modulated sub-carrier system', *CPEM Digest*, 1976, pp. 162–164

20 EVANS, D.S.: 'A simple digitizing system for humans in industry', *Instrum. Rev.*, 1966, 13, pp. 8–10 and 54–57

Determination of low values of attenuation from reflection coefficient measurements

Several papers[1-6] have described methods of deducing low values of attenuation from reflection coefficient measurements. These methods are discussed here.

9.1 Method using a standing-wave indicator

A popular method is to place the unknown between a standing-wave indicator and a sliding short circuit, with the input end of the unknown facing the short circuit, as shown in Fig. 9.1. The magnitudes and phase angles of the reflection coefficients are measured for several different positions of the short circuit and the results are plotted on polar graph paper. The points lie on a circle whose centre is usually displaced from the origin and it is shown later that the radius R, of this circle is given by

$$R = \frac{|s_{21}|^2}{1 - |s_{11}|^2} \tag{9.1}$$

Thus, after finding this radius, the dissipative component A_d of the attenuation through the unknown can be found immediately from eqn. 2.16, since

$$A_d = 10 \log_{10} \frac{1}{R} \tag{9.2}$$

A small correction may be needed to allow for the attenuation in the standing wave indicator and the sliding short circuit. To find the reflective component of the attenuation, the unknown is reversed, a perfectly matched load is connected to its output end, $|s_{11}|$ is measured and then eqn. 2.15 is used.

Eqn. 9.1 will now be derived. The signal flow graph of the unknown with the sliding short behind it is shown in Fig. 9.2. Using the non-touching loop rule, the reflection coefficient looking into the unknown from the standing wave detector is found to be

$$\Gamma_r = s_{22} + \frac{s_{12}s_{21}(-e^{-j2\beta l})}{1 - s_{11}(-e^{-j2\beta l})} \tag{9.3}$$

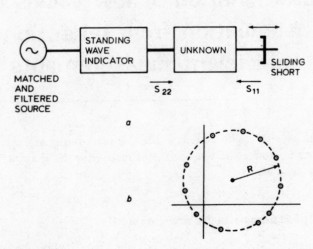

a

b

Fig. 9.1 Determination of low values of attenuation from reflection coefficient measurements
  a Equipment used
  b Polar plot of reflection coefficient for different positions of the sliding short

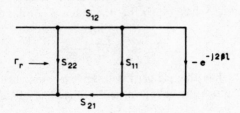

Fig. 9.2 *Signal flow graph of reversed* unknown with sliding short

where β is the phase constant and l is the distance from the input flange of the sliding short to the reflecting surface. Rearranging eqn. 9.3, we get

$$\Gamma_r = \frac{(s_{11}s_{22} - s_{12}s_{21})e^{-j2\beta l} + s_{22}}{s_{11}e^{-j2\beta l} + 1} \tag{9.4}$$

To proceed further, it is necessary to use the bilinear transformation (see Appendix 7). When

$$\Gamma_r = k\frac{ae^{j\theta} + b}{ce^{j\theta} + d} \qquad (9.5)$$

where a, b, c, d and k are complex quantities, it is shown in Appendix 7 that Γ_r traces out a circle in the complex plane as θ is varied. The centre of this circle is usually displaced from the origin. The radius R of this circle and the distance W from the origin to the centre are given by

$$R = |k|\frac{|bc - ad|}{|d|^2 - |c|^2} \qquad (9.6)$$

and

$$W = k\frac{bd^* - ac^*}{|d|^2 - |c|^2} \qquad (9.7)$$

On comparing eqn. 9.4 with eqn. 9.5 we see that

$$a = (s_{11}s_{22} - s_{12}s_{21}) \qquad d = 1$$
$$b = s_{22} \qquad\qquad k = 1$$
$$c = s_{11} \qquad\qquad \theta = -2\beta l$$

When these values are substituted into eqn. 9.6, we obtain eqn. 9.1, (since $s_{12} = s_{21}$ for any reciprocal 2-port).

This circle plotting technique is rather time consuming. If a slight reduction in accuracy is acceptable, a lot of work can be saved by simply measuring the maximum and minimum reflection coefficients, $|\Gamma|_{max}$ and $|\Gamma|_{min}$, that occur as the sliding short is moved along. We then have

$$|\Gamma|_{max} = R + |W| \qquad (9.8)$$

and

$$|\Gamma|_{min} = R - |W| \qquad (9.9)$$

Thus

$$R = \frac{|\Gamma|_{max} + |\Gamma|_{min}}{2} \qquad (9.10)$$

Some microwave workers have a strong preference for using standing wave ratios rather than reflection coefficients, so let S_{max} and S_{min} denote the maximum and minimum standing wave ratios that are obtained as the sliding short is moved along. Then, since

$$|\Gamma|_{max} = \frac{S_{max} - 1}{S_{max} + 1} \quad \text{and} \quad |\Gamma|_{min} = \frac{S_{min} - 1}{S_{min} + 1} \qquad \begin{array}{c}(9.11)\\(9.12)\end{array}$$

it immediately follows that

$$R = \frac{S_{max}S_{min} - 1}{(S_{max} + 1)(S_{min} + 1)}$$ (9.13)

Eqns. 9.8, 9.9, 9.10 and 9.13 apply to the usual case where the circle encloses the origin. If this is not so, obvious changes are needed to these equations.[3]

When this method is used to measure the attenuation in a short length of waveguide, the standing wave ratio is extremely high. Under these circumstances, improved accuracy can be obtained by measuring the standing wave ratio in a manner that was described by Roberts and Von Hippel,[7] in connection with determinations of dielectric constant and loss. This technique has been described in great detail in standard works.[8,9] Very briefly, it involves measuring the distance between two points, on either side of a minimum, at which the field strength is an arbitrary amount higher than the minimum field strength. Very frequently, this arbitrary amount is taken to be $\sqrt{2} = 3:01$ dB, and, if the measured distance with this assumption is denoted by w and the guide wavelength is λ_g, it is easily shown that the standing wave ratio (expressed as a quantity greater than unity) is given approximately by $\lambda_g/\pi w$. Thus, after measuring w, the only other task is to determine λ_g which is, of course, twice the distance between two adjacent minima. The simple formula quoted above is only valid for very high standing wave ratios. The standard works cited earlier[8,9] should be consulted for the full theory underlying this valuable technique. Owens[10] has described a technique that is even better than the one from Roberts and Von Hippel. This involves measuring two widths across the trough in the standing wave pattern at field strengths that are well above the noise level of the detection system.

P.J. Skilton[11] has carried out attenuation measurements at 7.3 GHz on numerous low-loss components, using this Roberts/Von Hippel technique with equipment similar to that shown in Fig. 9.1. His results have an uncertainty in the region of ± 0.003 dB for a confidence level of 99.5%.

9.2 Method using a reflectometer

Instead of using a slotted line, a reflectometer can be used. A detailed description will now be given of a reflectometer technique that has been widely used at the Royal Radar Establishment[12] for the measurement of low values of attenuation. This technique was developed to overcome a serious shortcoming of a conventional modulated sub-carrier attenuation measuring equipment. To measure the attenuation in a short length of waveguide with equipment such as that shown in Fig. 8.5, the two flanges at the insertion point must first of all be bolted directly together (to set the zero of the measuring equipment) and they then have to be moved apart and bolted to the two ends of the 'unknown'. To make this possible, it is necessary either to insert an extra piece of waveguide (equal in length to the unknown) in the carrier channel or have a length of flexible waveguide or flexible coaxial cable in one of the channels. The change over can take several minutes and during this time the zero can, under unfavourable circumstances, drift by a few thousandths of a decibel. Furthermore, the movements of the various coaxial and waveguide components and the mechanical strains produced by the unbolting and reconnecting of the flanges produce small changes in the transmitted signal. Our experience shows that if this complete process is repeated 10 times on the same unknown, the total spread in the 10 results can be as high as 0·01 dB.

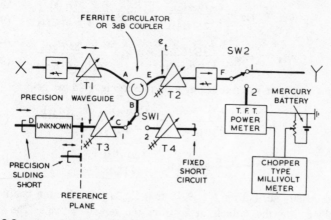

Fig. 9.3 Components inserted between X and Y in Fig. 8.5 when measuring low values of attenuation

This difficulty has been overcome by replacing the components between X and Y in Fig. 8.5 with the components shown in Fig. 9.3 which form a tuned reflectometer system[*] Under normal measurement conditions, the amplitude modulated signal follows the route XABCD-CBEFY and therefore passes through the unknown twice. The waveguide

[*] A ferrite circulator gives a more compact layout than a 3 dB coupler, but it can not always be tuned

switch SW1 is a most important item. By switching to the fixed short circuit, the zero of the modulated sub-carrier system can be reset rapidly at any time without unbolting or reconnecting any waveguide flanges.

The signal flow graph for the tuned reflectometer system is shown in Fig. 9.4a. Here, Γ_S represents the reflection coefficient looking into the source from A through T1 and Γ_t represents the reflection coefficient looking into the receiver from E through T2. The s-parameters of the path through the switch and tuner T3 are denoted by t_{11}, t_{12}, t_{21} and t_{22} and Γ_r is the reflection coefficient of the device that is connected to the reference plane.

The first stage of the tuning procedure is to adjust tuner T2 until a match is seen looking towards the receiver from E.

Using well known rules, the signal flow graph shown in Fig. 9.4a can now be simplified to the one in Fig. 9.4b where

$$n_{11} = m_{11} + \frac{m_{21} t_{11} m_{12}}{1 - m_{22} t_{11}} \qquad n_{12} = \frac{t_{12} m_{12}}{1 - m_{22} t_{11}} \qquad \begin{matrix} (9.14) \\ (9.15) \end{matrix}$$

$$n_{22} = t_{22} + \frac{t_{12} m_{22} t_{21}}{1 - m_{22} t_{11}} \qquad n_{21} = \frac{m_{21} t_{21}}{1 - m_{22} t_{11}} \qquad \begin{matrix} (9.16) \\ (9.17) \end{matrix}$$

$$n_{31} = m_{31} + \frac{m_{21} t_{11} m_{32}}{1 - m_{22} t_{11}} \qquad n_{32} = \frac{t_{12} m_{32}}{1 - m_{22} t_{11}} \qquad \begin{matrix} (9.18) \\ (9.19) \end{matrix}$$

Applying Mason's non-touching loop rule to Fig. 9.4b, we get

$$e_t = \frac{e_s}{1 - \Gamma_s n_{11}} \cdot \frac{n_{31} + \Gamma_r(n_{21} n_{32} - n_{31} n_{22})}{1 - \Gamma_r \Gamma_{2i}} \qquad (9.20)$$

where

$$\Gamma_{2i} = n_{22} + \frac{n_{12} \Gamma_s n_{21}}{1 - \Gamma_s n_{11}} \qquad (9.21)$$

It is readily seen from Fig. 9.4b that Γ_{2i} is the reflection coefficient looking into the reflectometer from the reference plane. The signal entering the receiver is made directly proportional to Γ_r by adjusting tuner T3 until $n_{31} = 0$ and then varying tuner T1 until $\Gamma_{2i} = 0$. The practical details of this tuning procedure are similar to those given by Anson[13] and are as follows:

(*a*) with a perfectly matched load connected to the reference plane and SW1 in position 1, T3 is adjusted until the reflectometer output signal e_t is zero

(*b*) with SW1 in position 2 and a perfectly matched load in place of the fixed short circuit, tuner T4 is adjusted until e_t is again zero

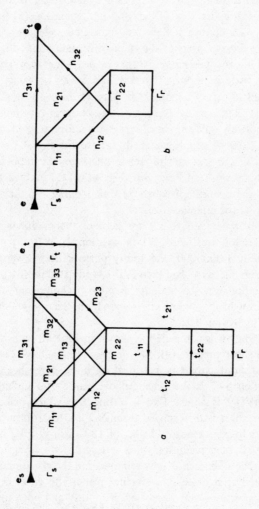

Fig. 9.4 Signal flow graphs
a Signal flow graph of the tuned reflectometer
b Simplified version of the signal flow graph shown in *a*

(*c*) with a precision sliding short circuit connected to the reference
plane and SW1 in position 1, T1 is adjusted until e_t stays constant as the short circuit is moved along.

Operation (*c*) is rather lengthy and is made easier by using, for T1,
a fully-screened sliding probe tuner rather than a 3 screw tuner.

A semi-automatic way of carrying out operation (*c*) has been described by Zanboorie.[14] The thin film thermoelectric power meter
is used for these 3 parts of the tuning procedure and also for part of
the measurement procedure. It has a head that can give full-scale
deflection with an input of $0.3\,\mu$W and, when necessary, its output
can be backed off by a mercury battery so that very small changes in
its output voltage can easily be seen on a chopper-type millivolt meter.

A completely different and very satisfactory way of carrying out
operation (*c*) is to couple matched loads to the Fig. 9.3 configuration
at X and Y, connect the reference plane to a computer-controlled
microwave network analyser and then adjust T1 until a good match
is seen, at the operating frequency, when looking into the tuned reflectometer from the reference plane.

The technique used to set the zero of the modulated sub-carrier
system is as follows: with SW1 in position 2, SW2 in position 1 and
IVD-2 set at 1.0000000, the rotary phase shifter is varied until V_p
reaches a suitable zero* and then IVD-1 is adjusted until V_A is zero.

The method used to find the dissipative component of the attenuation through the unknown is based on the two-point technique
described in Section 9.1.

The unknown is connected as shown in Fig. 9.3 with its input end
facing the short circuit. With SW1 in position 1 and SW2 in position 2,
the sliding short circuit is moved along until e_t reaches a maximum
value. Switch SW2 is then changed over and, after nulling the phase
channel, IVD-2 is adjusted to a setting D_1, which makes V_A zero.
The sliding short is then moved through $\lambda_g/4$, the rotary phase shifter
is adjusted to give a phase change of $180°$, bringing V_p back to zero
and IVD-2 is then readjusted to a setting D_2 which again brings V_A
to a null. The unknown is now removed and the precision sliding short
is connected straight to the reference plane. This is set in turn to the
same two positions that were used previously and both the phase and
amplitude channels are nulled in each case. Let D_3 and D_4 denote the
settings of IVD-2 that now make V_A zero. Then, with perfect tuning

---

* As the rotary phase shifter is turned through a complete circle, V_p will pass
through zero 4 times but only 2 of the settings will give the phase relationship
needed to obtain a balance in the amplitude channel. A correct setting can be
found in a few seconds by trial and error

of the reflectometer, it is proved indirectly later on in this chapter that the dissipative component of the attenuation through the unknown is given by

$$A_{dm} = 10 \log_{10} \frac{D_3 + D_4}{D_1 + D_2} \qquad (9.22)$$

When the tuners have been carefully adjusted, D_3 and D_4 differ from unity by only a few parts in 10.[4] The measurement procedure, has been carefully devised so that it is not necessary to make any correction for losses in the precision sliding short. The entire procedure could be carried out using only the modulated sub-carrier receiver but, after each adjustment, it would be necessary to null the phase channel, so it is much quicker to use the power meter wherever possible.

To give an idea of the remarkable repeatability which can be achieved with this method, Table 1 gives the mean values, random errors and total spreads obtained in measurements on 5 different WG11A components. Each one was bolted into position using dowels and shims, measured and removed again 10 times.

Table 1   Results of measurements on five WG11A components

Letter denoting waveguide configuration	Measured attenuation (mean value)	Standard deviation of the 10 results	Standard deviation of the mean	Total spread of the 10 results
	dB	dB	dB	dB
A	0·0746	0·00045	0·00014	0·0739–0·0752
B	0·0725	0·00040	0·00013	0·0717–0·0728
C	0·0844	0·00039	0·00012	0·0837–0·0849
D	0·0944	0·00043	0·00014	0·0938–0·0949
E	0·0909	0·00045	0·00014	0·0902–0·0912

Agreement to better than 0·001 dB has been obtained between this technique and the one described in Section 9.1, which involves the use of a standing wave indicator. However, the total spreads on 10 independent results are appreciably lower when this modulated sub-carrier method is used.

The theory underlying this method of measurement will now be derived. Substituting eqn. 9.3 into 9.20 we get

$$e_t = k' \frac{a'e^{j\theta} + b'}{c'e^{j\theta} + d'} \qquad (9.23)$$

where

$$a' = (n_{21}n_{32} - n_{31}n_{22})(s_{12}s_{21} - s_{11}s_{22}) - s_{11}n_{31} \qquad (9.24)$$

$$b' = n_{31} + s_{22}(n_{21}n_{32} - n_{31}n_{22}) \qquad (9.25)$$

$$c' = \{\Gamma_{2i}(s_{12}s_{21} - s_{11}s_{22}) + s_{11}\} \qquad (9.26)$$

$$d' = 1 - \Gamma_{2i}s_{22} \qquad (9.27)$$

$$k' = \frac{e_s}{1 - \Gamma_s n_{11}} \qquad (9.28)$$

$$\theta = -2\beta l. \qquad (9.29)$$

Using again the bilinear transformation expression given in eqn. 9.6, we now get

$$R = |k'| \frac{|s_{12}s_{21}| \cdot |n_{21}n_{32} + n_{31}\Gamma_{2i} - n_{31}n_{22}|}{|1 - \Gamma_{2i}s_{22}|^2 - |\Gamma_{2i}(s_{12}s_{21} - s_{11}s_{22}) + s_{11}|^2} \qquad (9.30)$$

When the unknown is removed and the sliding short is connected straight to the reference plane, the radius R_0 can be found from eqn. 9.30 by letting $s_{11} = s_{22} = 0$ and $s_{12} = s_{21} = 1$. Thus

$$R_0 = |k'| \frac{|n_{21}n_{32} + n_{31}\Gamma_{2i} - n_{31}n_{22}|}{1 - |\Gamma_{2i}|^2} \qquad (9.31)$$

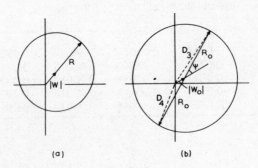

Fig. 9.5 Polar plots of e_t as l is varied
    a With unknown
    b Without unknown

The magnitude of the distance from the origin to the centre in this latter case is found from eqn. 9.7 to be

$$|W_0| = |k'| \frac{|n_{31} + (n_{21}n_{32} - n_{31}n_{22})\Gamma_{2i}^*|}{1 - |\Gamma_{2i}|^2} \qquad (9.32)$$

Thus

$$\frac{|W_0|}{R_0} = \frac{|n_{31} + (n_{21}n_{32} - n_{31}n_{22})\Gamma_{2i}^*|}{|n_{21}n_{32} + n_{31}\Gamma_{2i} - n_{31}n_{22}|} \qquad (9.33)$$

When n_{31} approaches zero:

$$\frac{|W_0|}{R_0} \approx |\Gamma_{2i}^*| = |\Gamma_{2i}| \qquad (9.34)$$

Eqn. 9.22 is used to calculate the dissipative component of the attenuation and, from the measurement procedure given earlier and Fig. 9.5, it follows that

$$D_1 = R + |W| \quad \text{and} \quad D_2 = R - |W| \qquad (9.35), (9.36)$$

Furthermore

$$D_3 + D_4 = (|W_0|^2 + R_0^2 + 2|W_0|R_0 \cos \psi)^{1/2} +$$
$$(|W_0|^2 + R_0^2 - 2|W_0|R_0 \cos \psi)^{1/2} \qquad (9.37)$$

$$\approx 2R_0 \text{ as } \frac{|W_0|}{R_0} \ll 1 \qquad (9.38)$$

Hence

$$A_{dm} = 10 \log_{10} \frac{R_0}{R} \qquad (9.39)$$

$$= 10 \log_{10} \frac{|1 - \Gamma_{2i}s_{22}|^2 - |\Gamma_{2i}(s_{12}s_{21} - s_{11}s_{22}) + s_{11}|^2}{|s_{12}s_{21}|(1 - |\Gamma_{2i}|^2)} \qquad (9.40)$$

When $\Gamma_{2i} = 0$, the dissipative component of the attenuation is given by

$$A_d = 10 \log_{10} \frac{1 - |s_{11}|^2}{|s_{12}s_{21}|} \qquad (9.41)$$

which is exactly correct (see eqn. 2.16).

If tuner T1 is not correctly set and Γ_{2i} is not zero, there is a small error in the attenuation measurement. This error is given by

$$\Delta A = A_{dm} - A_d = 10 \log_{10}$$
$$\frac{|1 - \Gamma_{2i}s_{22}|^2 - |\Gamma_{2i}(s_{12}s_{21} - s_{11}s_{22}) + s_{11}|^2}{(1 - |\Gamma_{2i}|^2)(1 - |s_{11}|^2)} \qquad (9.42)$$

Fig. 9.6 shows how the upper limit of this error varies with $|s_{11}|$ for different values of Γ_{2i} and A, when the unknown is symmetrical.

If, at the end of stage (*c*) of the tuning procedure, e_t has maximum and minimum values of e_{max} and e_{min}, then it follows from eqn. 9.34 that

$$e_{max} = R_0 + R_0 |\Gamma_{2i}| \qquad (9.43)$$

and

$$e_{min} = R_0 - R_0 |\Gamma_{2i}| \qquad (9.44)$$

Hence

$$\frac{e_{max}}{e_{min}} = \frac{1 + |\Gamma_{2i}|}{1 - |\Gamma_{2i}|} \approx 1 + 2 |\Gamma_{2i}| \qquad (9.45)$$

as $|\Gamma_2|$ is very small. Rearranging eqn. 9.45, we get

$$|\Gamma_{2i}| = \frac{1}{2} \left\{ \frac{e_{max}}{e_{min}} - 1 \right\} \qquad (9.46)$$

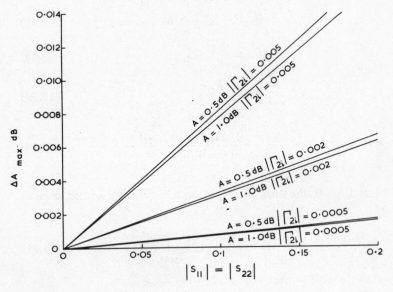

Fig. 9.6 Curves showing the error ΔA_{max} in the attenuation measurement versus $|s_{11}|$ for various values of *A* and $|\Gamma_{2i}|$

Thus $|\Gamma_{2i}|$ can easily be found by simply measuring e_{max} and e_{min} at the end of the tuning procedure. If e_{max}/e_{min} is reduced to less than 1·001 and the unknown has a voltage reflection coefficient of less than 0·02, it is seen from Fig. 9.6 that ΔA is less than 0·0002 dB.

If the attenuation in the sliding short is appreciable, the curve of e_t versus *l* will have a general slope downwards as shown in Fig. 9.7 and it is then necessary to replace

$$\frac{e_{max}}{e_{min}} \quad \text{in eqn. 9.46 by} \quad \frac{e_{max,1} + e_{max,2}}{2e_{min}}$$

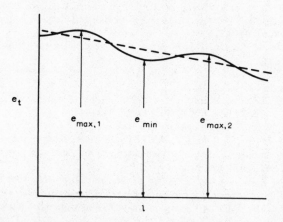

Fig. 9.7 Diagram showing the measurements needed to determine $|\Gamma_{2i}|$ when the attenuation in the sliding short is appreciable

9.3 Conclusion

The reflection coefficient methods of attenuation measurement that have been described in this chapter are capable of giving highly accurate values of attenuation lying in the range 0 to 1 dB. Furthermore, if the method described in Section 9.1 is used, only a small amount of test equipment is required. All of the items that are needed are likely to be available in almost any microwave laboratory.

9.4 References

1 BLACKBAND, W.T. and BROWN, D.R.: 'The two-point method of measuring characteristic impedance and attenuation of cables at 3000 Mc/s', *J. IEE*, 1946, **93**, Pt. IIIA, pp. 1383–1386

2 CULLEN, A.L.: 'Measurement of microwave transmission efficiency', *Wireless Eng.*, 1949, **26**, pp. 255–257

3 BEATTY, R.W.: 'Determination of attenuation from impedance measurements', *Proc. IRE*, 1950, **38**, pp. 895–897

4 DESCHAMPS, G.A.: 'Determination of reflection coefficients and insertion loss of a wave-guide junction', *J. Appl. Phys.*, 1953, **24**, pp. 1046–1050

5 VOGELMAN, J.H.: 'Precision measurement of waveguide attenuation', *Electronics*, 1953, **26**, pp. 196–199

6 POMEROY, A.F., and SUAREZ, E.M.: 'Determining attenuation of wave-guide from electrical measurements on short samples', *IRE Trans.*, 1956, **MTT-4**, pp. 122–129

7 ROBERTS, S., and VON HIPPEL, A.: 'A new method for measuring dielectric constant and loss in the range of centimetric waves', *J. Appl. Phys.*, **1946, 17**, pp. 610–616

8 BARLOW, H.M. and CULLEN, A.L.: 'Micro-wave measurements' (Constable, 1950), Chap. 5

9 SUCHER, M., and FOX, J.: 'Handbook of microwave measurements – Vol. 1' (Polytechnic Press, 1963), Chap. 2

10 OWENS, R.P.: 'Technique for the measurement of standing-wave ratios at low power level', *Proc. IEE*, 1969, **116**, pp. 933–940

11 SKILTON, P.J.: Unpublished MOD(PE) work

12 WARNER, F.L., HERMAN, P. and JEFFS, T.: 'Special techniques for measuring low and high values of attenuation with a modulated sub-carrier system', *IEEE Trans.*, 1974, **IM-23**, pp. 381–386

13 ANSON, W.J.: 'A guide to the use of the modified reflectometer technique of VSWR measurement', *J. Res. NBS*, 1961, **65C**, pp. 217–223

14 ZANBOORIE, M.H.: 'A semi-automatic technique for tuning a reflectometer', *IEEE Trans.*, 1965, **MTT-13**, pp. 709–710

Chapter 10

Swept frequency techniques

10.1 Introduction

Before 1957, broadband frequency response measurements on microwave components were nearly always carried out using laborious point-by-point techniques. The commercial introduction of electronically tunable backward wave oscillators in 1957 drastically changed this situation and, since then, swept frequency measurement systems have come into widespread use. They save a tremendous amount of time, resonances which can easily be missed using point-by-point methods are clearly revealed and the effects of adjustments made to the device under test can be seen immediately over a wide band.

Various components that are widely used in swept frequency measuring systems are described in Section 10.2 and several different configurations that can be used to carry out swept frequency attenuation measurements are discussed in Section 10.3.

10.2 Components for swept frequency systems

10.2.1 Swept sources

A reflex klystron[1] can be tuned electronically by varying the voltage applied to its reflector. The tuning range that can be obtained in this way is quite small e.g. with a CV2346 reflex klystron operating at 10 GHz, it is typically 20 MHz between the half power points. This is adequate for making swept frequency measurements on high Q cavities, but electronic tuning ranges in the region of one octave are needed to carry out full tests on broadband waveguide components.

In two early broadband swept frequency measuring equipments described by Hunton and Pappas,[2] reflex klystrons were used as the sources and these were tuned mechanically, over the bands 2·6–3·95 GHz and 8·2–10 GHz by geared-down electric motors. When this technique is used, the reflector voltage potentiometer must be suitably ganged to the mechanical tuner, as the optimum reflector voltage varies considerably with frequency. This solution is therefore quite cumbersome. However, it has also been used by Kinnear.[3]

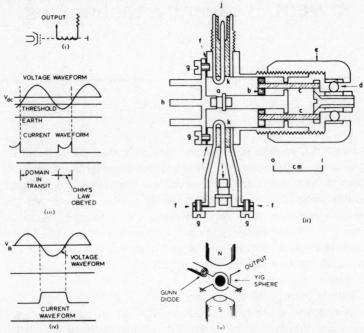

Fig. 10.1 Microwave oscillators
i   Backward wave oscillator
ii Gunn oscillator with both mechanical and electronic tuning, showing:

a Gunn diode	*e* tuning knob	*i* varactor
b alumina washer	*f* mica	*j* output connector
c dielectric rods	*g* nylon screws	*k* rotatable coupling loops
d ball race	*h* copper heat sink	

iii Voltage and current waveforms of a Gunn oscillator
iv Voltage and current waveforms of an impatt oscillator
v YIG tuned Gunn oscillator.

As mentioned earlier, a major break-through in the swept frequency measurement field occurred in 1957 when electronically-tunable backward wave oscillators became commercially available. A simplified diagram of a backward wave oscillator (BWO)[4] is shown in Fig. 10.1(*i*). An electron beam is passed through a helix that has a steady axial magnetic field applied to it. Interaction occurs between the electron stream and a backward wave on the helix, and the tube oscillates at a frequency proportional to the square root of the cathode helix voltage. Thus, a 4 to 1 variation in this voltage gives electronic tuning over an octave.

Lacy and Wheeler[5] described the first fully-engineered swept source of this type. It could be tuned linearly over the range 8·2 to 12·4 GHz by varying the helix voltage and the sweep time could be varied from approximately 0·01 second to 100 seconds. The output power exceeded 10 mW over the entire band.

BWO swept sources were developed to a very high degree in the next 6 years[6] and they are now available from several manufacturers. By using a main frame and 10 different plug-in BWOs, it is now possible to obtain coverage from 1 to 110 GHz. Table 10.1 gives typical values for the frequency ranges and output powers that are currently obtainable with BWOs. By using highly stabilised power units, the residual frequency modulation can be reduced, typically, to 2 p.p.m. at centimetre wavelengths, and 5 p.p.m. at millimetre wavelengths. A frequency sweep that is linear with time to within about 1% is achieved by generating a sawtooth waveform and then shaping it with a network containing many diodes and resistors to obtain a close approximation to the required waveform. Associated circuits provide frequency markers and several sweep modes, and also allow the BWO to be amplitude modulated by internal or external signals.

If the output from a swept source is inadequate, it can be followed by a travelling-wave tube power amplifier.[7] When this is done, modulation waveforms can be applied to the amplifier[8] and the frequency can be changed by applying a sawtooth waveform of the correct amplitude to the helix in it.[9] When a TWT power amplifier is used, care is needed to make sure that no components are damaged by the high power level.

The power outputs from currently available millimetre wave swept sources are rather low (see Table 10.1) but millimetre wave BWOs with outputs in excess of 1 watt up to 300 GHz have been available for several years.[10]   Although BWOs are now widely used, their prices have stayed high and, being hot cathode tubes, their lives are quite short. BWOs that are used throughout each working day rarely last for more than 2 years. The cost of replacement BWOs can form a high percentage of the annual cost of running a microwave laboratory. Great efforts have therefore been made in recent years to develop solid-state swept sources. Both bipolar and field-effect transistors have now been developed to the stage where they will give very useful output powers[11] up to about 5 GHz and the remaining part of the microwave spectrum can now be covered by Gunn[12,13] and Impatt[14] oscillators.

Table 10.1   Output power of swept sources

Frequency range GHz	Minimum output power over band with a BWO	Minimum output power over band with a solid-state source
1–2	100 mW	30 mW
2–4	85	38
4–8	40	40
8–12·4	60	30
12·4–18	40	27
18–26·5	25	10
26·5–40	5	
40–60	3	
60–90	3	
90–120	1	

A typical Gunn oscillator is shown in Fig. 10.1(*ii*). The Gunn diode contains an *n*-type GaAs epitaxial layer, of thickness *t*, on a high conductivity GaAs substrate. A grossly simplified description of the popular delayed domain mode of operation will now be given.

When the applied field exceeds a threshold value of 3000 V/cm, the electrons near the cathode are forced into an energy level where the mobility is about 70 times lower. This causes the formation of a domain, which travels to the anode at a velocity v ($\approx 10^7$ cm/s) and then collapses, While the domain is in transit, the current stays almost constant at approximately half its threshold value. After the collapse of the domain, the current varies in accordance with Ohm's law until the field again reaches 3000 V/cm. This pulsating current causes the cavity to oscillate at its resonant frequency. The idealised voltage and current waveforms are shown in Fig. 10.1(*iii*). The oscillation frequency can be varied from approximately $v/2t$ to v/t by tuning the cavity. If $t = 10 \mu m$, the threshold is reached at about 3 volts and the tuning range is approximately 5 to 10GHz.

An impatt oscillator consists of a n^+pip^+ diode inside a resonant cavity. The diode is reverse-biased to the avalanche breakdown voltage V_B. It then looks like a negative resistance and maintains an oscillation at the cavity resonant frequency. An avalanche builds up throughout each positive half cycle at the n^+p boundary, where the electric field is highest. On the negative halfcycles, the voltage falls below V_B and the whole process decays. The electrons produced in the avalanche are swept immediately into the n^+ contact while the holes are injected into the intrinsic region, of width W, and drift across it at a saturation velocity v_s ($\approx 10^7$ cm/s). While the holes are in transit, a current is induced in the external circuit and this is in antiphase with the r.f. voltage. Oscillations occur when the resonant frequency of the cavity is close to $v_s/2W$. The voltage and current waveforms of an impatt oscillator are shown in Fig. 10.1(iv).

Solid state oscillators are very small, they operate from low voltages and they should have much longer lives than BWOs.

Several different methods of tuning solid-state oscillators electronically have been investigated. Variation of permeability,[15] variation of permittivity and piezoelectric methods have not been very successful and the two best techniques that have been found are:

(a) variation of the reverse bias on a varactor which is coupled into the oscillator cavity[16-42] [see Fig. 10.1(ii)]

(b) variation of the magnetic field applied to an yttrium iron garnet (YIG) sphere which forms the tuned circuit.[43-51]

With varactor tuning, a shaping circuit is required to obtain a linear sweep, a high cut-off frequency varactor is needed to avoid a serious reduction in the output power, and harmonics are generated by the varactor. However, extremely rapid tuning rates can be obtained with this technique, and, with very careful design, octave bandwidths can be achieved. Furthermore, no power is needed to produce the tuning.

In a YIG tuned Gunn oscillator, orthogonal loops [see Fig. 10.1(v)] are used to couple the Gunn diode and load to the YIG sphere, which usually has a diameter of less than 1 mm. The ferromagnetic resonant frequency of a YIG sphere is given in Hertz by

$$f = 3 \cdot 5 \times 10^4 \, H \tag{10.1}$$

where H is the applied magnetic field in Am^{-1}. Thus, there is an almost linear relationship between the oscillation frequency and the current through the electromagnet. In practice, the departure from linearity can be made less than 0·1%. The electromagnet makes a YIG-tuned oscillator much heavier than a varactor tuned one and the sweep rate is limited to about 1 GHz/ms by the magnetic driving circuit. At 18 GHz, about 10 watts are dissipated in the tuning coil. The temperature coefficient of a YIG-tuned oscillator can be made very low by careful orientation of the YIG sphere relative to the external field. A YIG sphere has an unloaded Q of about 10 000 and the spectral width of the output from a YIG-tuned oscillator is as narrow as that from a reflex klystron. Saturation effects in the YIG appear to limit the output power to a value somewhere in the range 10 to 100 mW. The Gunn diode limits the electronic tuning range of a YIG-Gunn oscillator to about 1 octave. Eddy currents in the pole pieces cause the frequency to lag behind the tuning current by an amount which depends on the sweep rate. Compensation circuits have been developed to overcome this trouble.[52] Very low ripple is essential in the current supply to keep frequency modulation down to a few p.p.m.

Impatt oscillators tend to give higher output power than Gunn oscillators, but they are noisier and more difficult to tune, so they have not been used much in swept sources. However, several manufacturers now offer solid-state swept sources that contain YIG-tuned transistor oscillators for use up to about 4 GHz, and YIG-tuned Gunn oscillators for use in the various bands between 4 GHz and 26·5 GHz. Typical values for the minimum output powers from these solid-state sources at the time of writing (March 1975) are given in Table 10.1.

Multi-band swept sources that will, for example, cover the entire band 0·1 to 18 GHz have been developed in recent years.[53] These contain several conventional swept sources that are selected in rapid succession by *p-i-n* diode switches.

10.2.2 Broadband detectors

Microwave detector diodes[54, 55] are used in many swept frequency measuring equipments. A microwave diode, by itself, presents a poor match to a 50 Ω transmission line. An early wideband coaxial crystal detector described by Schrock[56] had variations of approximately ± 4 dB in its response over the frequency range 10 MHz to 12·4 GHz and a maximum v.s.w.r. of 3 over this range. A greatly improved co-axial crystal detector was described by Riley[57] in 1963. The coaxial line is terminated by a 50 Ω disc resistor and a small diode is mounted immediately behind it. These two components and a 10pF bypass capacitor are mounted together, in a small replaceable capsule. Up to 12·4 GHz, this combination gives a frequency response that is flat to within ± 0·5 dB and a v.s.w.r. below 1·5. Although a lot of the input power is absorbed by the disc resistor, a low level sensitivity of > 0·4 mV/μW is achieved. Coaxial detectors with performance specifications similar to this can now be supplied by several firms. A range of wave-guide crystal detectors with small frequency response variations was described by Prickett and Reniham[58] in 1965.

A microwave diode acts as a square law detector (output voltage proportional to input power) at low input levels and as a linear detector (output voltage proportional to input voltage) when it is driven very hard. Many measurement applications require a wide square law region. This region can be maximised by careful choice of the load resistance, a value between 2 and $3\,k\Omega$ frequently gives best results. When a detector of the type described by Riley[57] has been optimised in this manner, it retains its square law characteristic to within $\pm 0.5\,dB$ up to an output of 50 mV.

A forward-biased* Schottky barrier diode for use in swept frequency measurement systems has been described recently by David.[59] This has an r.f. bandwidth of 0·1 to 18 GHz and a dynamic range of 60 dB with an amplifier bandwidth of 30 kHz. Due to the use of forward bias, it is only suitable for use with modulated input signals. Diodes of this type are manufactured using thin-film hybrid microcircuit techniques and the variation in characteristics from detector to detector is small, e.g. $< 0.25\,dB$ over the entire dynamic range, and < 1.25 dB over the full frequency range. A good match is achieved by shunting the detector with a $57\,\Omega$ resistor.

Barretters (see Chapter 8 and Appendix 5) can be used instead of diode detectors in swept frequency measurement systems. However, their slow response time seriously limits the sweep speed and necessitates the use of an X-Y recorder rather than an oscilloscope, to display the results.

10.2.3 Directional Couplers

Directional couplers are widely used in swept frequency measurement systems. For this role, the requirements are: almost constant coupling, high directivity and good matching over a wide frequency range. The early directional couplers[60] had many shortcomings. However, a big advance was made in 1952 by Barnett and Hunton,[61] who developed very wideband 10 and 20 dB waveguide directional couplers with much higher directivities than were hitherto possible. They accomplished this with a large number of coupling holes whose diameters and positions were chosen to give a Tschebyscheff (equal ripple) response over the required band. A year later, Barnett[62] reported similar achievements with a 3 dB waveguide coupler. High performance waveguide directional couplers of this type are now commercially available in all popular waveguide sizes from many different firms; a typical specification being as follows: coupling variation $< \pm 0.5\,dB$, directivity $>$ 40 dB and v.s.w.r. < 1.05 over the entire band covered by the waveguide.

---

* The theory of detection with forward bias has been given by Warner[55]

Progress on coaxial directional couplers was less rapid. In 1955, Hunton, Poulter and Reis[63] described a range of octave bandwidth coaxial directional couplers with coupling variations of ± 1 dB and directivities of 26 to 30 dB. An improved range of coaxial directional couplers was described by Prickett[64] in 1965, and two years later Anderson and Dennison[65] announced the development of a coaxial directional coupler which covers four octaves and has a directivity that is > 26 dB through this enormous band and > 40 dB over a large part of it.

10.2.4 P-I-N diode levellers

P-I-N diode microwave amplitude stabilisers[66,67] have already been mentioned in Chapters 5 and 8. They are used extensively in swept frequency measurement systems because the output power from a swept source may vary by more than 10 dB as it is tuned over an octave. Besides stabilising the output power, a *p-i-n* diode leveller makes the source look well matched throughout the operating frequency range. This is an important feature, so it will be analysed mathematically. A *p-i-n* diode leveller is shown in Fig. 10.2*a* and the signal flow graph for it is shown in Fig. 10.2*b*, where Γ_G represents the reflection coefficient looking back into the *p-i-n* diode.

Using the non-touching loop rule, we find at once that

$$\frac{b_3}{e} = \frac{s_{31}(1 - s_{22}\Gamma_L) + s_{21}\Gamma_L s_{32}}{D} \tag{10.2}$$

and

$$\frac{b_2}{e} = \frac{s_{21}(1 - s_{33}\Gamma_D) + s_{31}\Gamma_D s_{23}}{D} \tag{10.3}$$

where

$D = 1 -$ sum of all 1st-order loops $+$ sum of all 2nd-order loops etc.

Also

$$a_2 = b_2\Gamma_L \tag{10.4}$$

Eliminating D, e and Γ_L from eqns. 10.2, 10.3 and 10.4, we get

$$b_2 = b_3\left\{\frac{s_{21}}{s_{31}} + \Gamma_D\left(s_{23} - \frac{s_{21}s_{33}}{s_{31}}\right)\right\} + a_2\left(s_{22} - \frac{s_{21}s_{32}}{s_{31}}\right) \tag{10.5}$$

The feedback loop keeps b_3 constant. If the entire levelled source is represented by a simple generator with a reflection coefficient Γ_G'', we have the signal flow graph shown in Fig. 10.2*c* and it is seen at once that

$$b_2 = e'' + a_2\Gamma_G'' \tag{10.6}$$

On comparing eqns. 10.5 and 10.6 we see that the reflection coefficient looking back into the *p-i-n* diode leveller is

$$\Gamma_G'' = s_{22} - \frac{s_{21} s_{32}}{s_{31}} \tag{10.7}$$

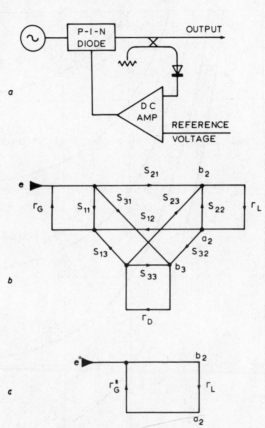

Fig. 10.2 *P-I-N* diode leveller and its signal flow graphs

The directivity of the directional coupler in decibels is given by

$$d = 20 \log_{10} \left| \frac{s_{31}}{s_{32}} \right| \tag{10.8}$$

and s_{21} is close to unity, so

$$|\Gamma_G''|_{max} \approx |s_{22}| + 10^{-(d/20)} \tag{10.9}$$

Thus, to obtain a good source match, we want $|s_{22}|$ to be as low as possible and the directivity to be as high as possible. Fig. 10.3 shows how the upper limit of the effective source reflection coefficient varies with directivity for different values of $|s_{22}|$.

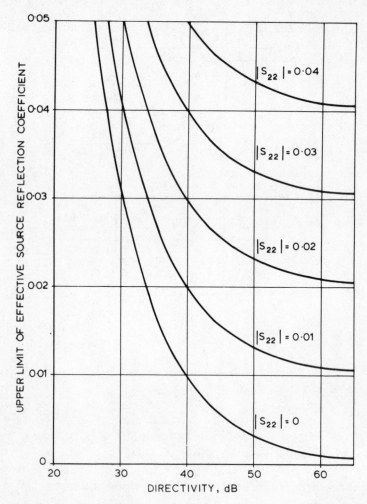

Fig. 10.3 Variation of effective source reflection coefficient with coupler directivity when using a *p-i-n* diode levelling loop

10.2.5 Ratiometers

A ratiometer often forms an important part of a swept frequency measurement system. It is an electronic instrument that is not particularly well known, so it will be briefly described. A ratiometer accepts two audio-frequency voltages and gives an output which is proportional to their ratio, the square root of their ratio or the logarithm of their ratio. An early ratiometer was described by Pappas.[68] A modern one has been described by Hoigaard[69] and a simplified block diagram of it is shown in Fig. 10.4. The two input signals are passed through quite separate channels that both contain a low-noise variable-gain tuned amplifier, a linear detector, a lowpass filter and a logarithmic amplifier. The outputs from the two logarithmic amplifiers are fed to a differential amplifier. This gives an output proportional to the logarithm of the ratio of the two input signals (see Fig. 10.4). If the two input signals are derived from a common source, variations in the amplitude of this source do not affect the reading on the output meter. In another recent ratiometer described by Vifian *et al.*[59] the logarithmic amplifiers precede the linear detectors. This modification compresses the dynamic range of the signals that have to be linearly detected and enables correct ·operation to be obtained when the two input voltages differ by 120 dB.

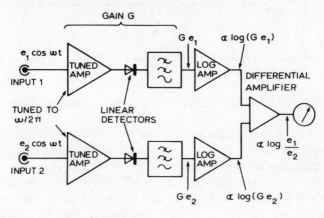

Fig. 10.4 Block diagram of ratiometer

10.3 Swept frequency attenuation measuring equipment

Two of the earliest swept frequency measuring systems.[2,63] were designed to display only the modulus of the reflection coefficient of the unknown against frequency. These equipments therefore fall outside the scope of this book.

The first swept frequency attenuation measuring equipment appears to have been described by Lacy and Wheeler,[5] and a block diagram of it is shown in Fig. 10.5. The output from the swept source is square-wave modulated at 1 kHz. Part of the incident power is tapped off by a broadband directional coupler and fed into a broadband matched diode detector. The remainder of the incident power enters the unknown and part of it emerges from the far side. The matched load absorbs most of this transmitted power, but a fraction of it is tapped off by the second directional coupler and fed into the second diode detector. Both the directional couplers and the diode detectors should

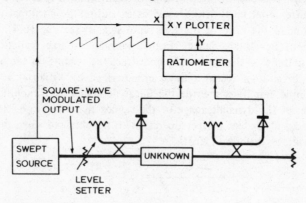

Fig. 10.5 Swept frequency attenuation measuring system described by Lacy and Wheeler in 1957

be 'matched pairs'. Their imperfections will then almost cancel each other out. The input power level is adjusted by the variable attenuator so that both diode detectors operate well within their square law regions. The outputs from the diode detectors are applied to the ratio-meter and this displays the ratio of the two tapped-off power levels directly in decibels. This ratio is not affected by the output power variations that occur as the frequency is swept. The X-Y plotter provides a permanent record of the attenuation through the unknown versus frequency. The output from the ratiometer is applied to its Y input and a sawtooth waveform, whose amplitude is linearly related to the frequency, is applied to its X input. The tuned amplifiers in the ratiometer have narrow bandwidths, so a slow sweep rate must be used to obtain the true shape of the attenuation versus frequency curve. The base line is established by plotting a curve with the two flanges at the insertion point joined directly together. If the variable attenuator is well matched and is set to provide at least 10 dB of attenuation, a fairly good match will be seen looking towards the source from the insertion point and the load will provide a good match looking in the other direction.

In a slightly later paper by Lacy and Wheeler,[70] a swept frequency measuring system with an oscilloscope display is described. The square-wave modulator, the first directional coupler, the first diode detector, the ratiometer and the X-Y plotter are no longer used and a d.c. amplifier with a response extending up to 300 kHz is used to amplify the output from the remaining diode detector. With this arrangement, fast sweeps can be used and the effects of adjustments to the device under test can be seen immediately over the entire swept band on the oscilloscope. However, the display is now confused by the power variations that occur in the output from the BWO as its frequency is swept. Furthermore, the dynamic range is reduced by more than 20 dB, because the drift plus the $1/f$ noise of a d.c. amplifier is many times greater than the noise originating in the narrow-band 1 kHz tuned amplifiers that are used in a ratiometer.

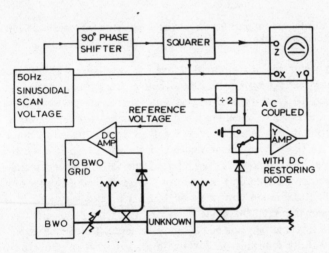

Fig. 10.6 Swept frequency attenuation measuring system described by Dix and Sherry in 1959

In 1959, Dix and Sherry[71] used a negative feedback system to eliminate the power variations from the source. A simplified diagram of their system is shown in Fig. 10.6. The output from the first diode detector is compared with a stabilised reference voltage and any small difference that occurs between these two voltages is amplified and then applied to the grid of the BWO. This changes the output power in a direction which tends to keep the diode detector output voltage almost exactly equal to the stabilised reference voltage throughout each sweep.

The vibrating switch grounds the Y amplifier input on alternate sweeps. This gives a base line on the display and enables an a.c. coupled Y amplifier, with a d.c. restoring diode, to be used.

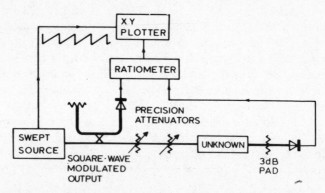

Fig. 10.7 Equipment used for the calibration grid technique described by Hunton and Lorence in 1960

A major advance in swept frequency techniques was described by Hunton and Lorence[72] in 1960. Their configuration, which is shown in Fig. 10.7, contains two precision variable attenuators. These are used to establish a calibration grid. With the two connectors at the insertion point joined directly together, these attenuators are set in turn to several convenient values and a single sweep is made at each setting. The unknown is then inserted, the variable attenuators are set at zero and a recording is made on the same sheet of paper over the existing calibration grid (see Fig. 10.8). With this technique, it is not necessary to keep the diode detectors within their square law regions and a dynamic range of 50 dB can be achieved. This can be extended[73] to 70 dB by following the BWO with a 1 watt TWT amplifier, provided that the 'unknown' will stand this power level without undergoing any change in its characteristics.

The next advance in swept frequency measurement techniques was made when *p-i-n* diode electronically-variable microwave attenuators became available. The amplitude stabilisation system described by Dix and Sherry[71] has a disadvantage. The amplified error voltage that is applied to the grid of the BWO changes the frequency as well as the output power and this slightly upsets the required linear frequency sweep. In 1963, Ely and Dudley[67] overcame this shortcoming by applying the amplified error voltage to a *p-i-n* diode instead of to the grid of the BWO. Fig. 10.9 shows a block diagram of the swept frequency at-

tenuation measuring system that they described. Systems of this type
have been widely used in the last decade. The *p-i-n* diode leveller removes the source power variations, keeps the source reflection coefficient low throughout the swept band, eliminates the need for a
ratiometer and enables an almost instantaneous display to be provided
on an oscilloscope. By using a logarithmic amplifier, the attenuation
can be displayed directly in decibels against frequency on the display.[74]
The dynamic range is limited to about 30 dB with such a system, the
lower limit being set by drift and $1/f$ noise in the d.c. amplifier and the
upper limit being determined by departure from the required square
law characteristic in the second diode detector. With a system of this
type, the accuracy is typically ± 0.5 dB when waveguide components
are used and ± 0.8 dB when coaxial components are used.[74] The errors
arise from:

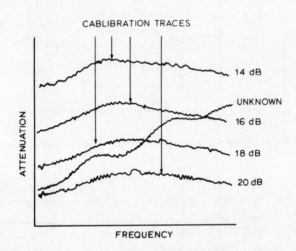

Fig. 10.8 Typical recording obtained with the equipment shown in Fig. 10.7

(*a*) different frequency responses in the two diode detectors
(*b*) frequency dependent differences between the coupling co-
    efficients of the two directional couplers
(*c*) incomplete removal of the source power variations by the *p-i-n*
    diode leveller
(*d*) imperfect matching looking in each direction from the insertion
    point
(*e*) deviations from square law in the diode detectors
(*f*) leakage
(*g*) noise and drift.

In spite of the popularity of the configuration shown in Fig. 10.9, interest in ratiometer methods has continued.[59, 69, 75] In 1965, Britton[75] described a ratiometer system with a dynamic range of 60 dB. This large range was achieved by allowing the diode detectors to operate almost up to the reverse breakdown level and compensation circuits were placed in the amplifiers to correct for the non-square-law behaviour. In this system, an unmodulated swept microwave signal, lowdrift d.c. amplifiers and a d.c. ratiometer are used. The drift has to be corrected manually a few times per hour.

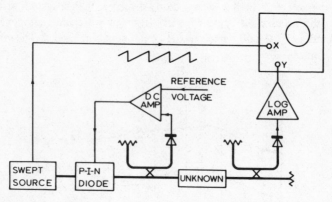

Fig. 10.9 Swept frequency attenuation measuring system containing a *p-i-n* diode leveller

Recently, Vifian, David and Frederick[59] have achieved a dynamic range of 60 dB with a modulated ratiometer system. In the early ratiometer systems, a modulation frequency of 1 kHz was normally used. Vifian and his colleagues raised this to 30 kHz, placed 30 kHz bandpass filters in the ratiometer, used the latest Schottky barrier detector diodes with forward bias, operated them beyond the square law region and placed compensation circuits in the amplifiers. By using this high modulation frequency, the $1/f$ noise from the detectors and amplifiers is reduced to a very low level and the bandwidths of the tuned amplifiers can be made large enough to give a flicker-free oscilloscope display. (With the early ratiometers, an X-Y recorder had to be used due to the narrow bandwidths of the tuned amplifiers). With a few specially-developed ultra-broadband components, this system can be used throughout the entire range from 100 MHz to 18 GHz and the drift is typically less than 0·2 dB over a 12 hour period without any correction.

Another major step forward in the swept frequency measurement field was described by B.O. Weinschel[76] in 1969. With the calibration grid technique described earlier (see Fig. 10.7), the calibration lines are curved as shown in Fig. 10.8 and this makes interpretation of the results more difficult. To overcome this shortcoming, Weinschel and his colleagues devised a lineariser which automatically converts the curved calibration lines to straight ones. A two-track endless magnetic tape is used. One track is used to control the frequency of the swept source. With the two connectors at the insertion point joined directly together and the two standard attenuators set at zero, the output from the ratiometer is stored on the other track of the magnetic tape. Measurements are then made at the other standard attenuator settings and the stored information is subtracted from the ratiometer output throughout each sweep before it is applied to the *X-Y* plotter. Since all of the calibration lines go up and down together with the system described by Hunton and Lorence,[72] they are all converted to straight lines by this new linearisation technique. The same procedure is then repeated after the unknown has been inserted.

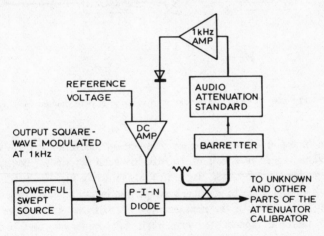

Fig. 10.10 Weinschel's technique for extending the range of attenuation measurements

A slightly different linearisation technique has been described by Britton.[77] The principles are the same, but, instead of using an endless magnetic tape, he used a small semiconductor memory to store the correction voltages.

In 1972, Weinschel[78] described a new technique for extending the range of attenuation measurements. This is shown in Fig. 10.10. A powerful source is needed. When there is high attenuation through the unknown, the power delivered by the levelling loop can be increased by almost exactly 10, 20 or 30 dB by changing the audio standard attenuator in the feedback loop. In most *p-i-n* diode levellers, a d.c. amplifier is used, but in this case, to keep the noise low, a 1 kHz tuned amplifier is used and the output from the source is square wave modulated at 1 kHz. Comparison with the d.c. reference voltage occurs after the second detector. A very low source reflection coefficient is maintained under all conditions, as explained in Section 10.2.4. When this system is used, a suitable sweep speed is about 25 seconds per octave, so the attenuation results must be displayed on an *X-Y* recorder.

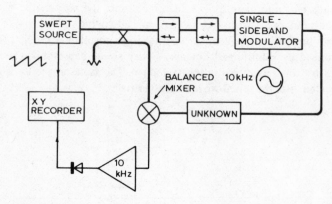

Fig. 10.11 Swept frequency attenuation measuring system devised by D.E. De Jersey

An ingenious swept frequency attenuation measuring system, with a dynamic range in excess of 100 dB, has been described recently by D.E. de Jersey.[79] A block diagram of it is shown in Fig. 10.11. It bears a close resemblance to Kinnear's swept frequency impedance meter,[3] which is described in the next chapter. Part of the output from the swept source is tapped off by a directional coupler and used to provide the local oscillator drive for a balanced mixer. The remaining power is passed through two isolators into a ferrite single-sideband modulator where its frequency is changed by 10 kHz. It is then fed through the attenuator under test into the mixer. An intermediate frequency signal of 10 kHz is therefore produced and this is passed through a logarithmic amplifier with a bandwidth of 1 kHz, detected and then fed to an *X-Y* plotter.

The stability of the intermediate frequency is independent of the frequency stability of the swept source and the bandwidth of the tuned amplifier can be made much narrower than 1 kHz if an exceptionally large dynamic range is needed. When the source power is 10 mW, the loss in the signal channel (excluding the unknown) is 4 dB, the noise factor is 20 dB and the amplifier bandwidth is 1 kHz, the signal power becomes equal to the total noise power referred to the mixer input when the attenuation through the device under test is 130 dB. This enormous dynamic range is achieved because the microwave diodes are used as mixers rather than detectors.

The single-sideband modulator has a bandwidth of 6% and a conversion loss of 1 to 2 dB. The two isolators are needed to prevent a reflected component at the sideband frequency reaching the mixer through a second reflection from the source.

In some small firms and colleges, electronically-swept sources may not be available. However, any of the swept frequency techniques described in this chapter that use X-Y recorders can be used with a series of low-cost manually-tuned signal sources that have been described recently by Draper and Hurst.[80] In these sources, the waveform required by the X amplifier in the recorder is generated by a potentiometer that is mechanically coupled to the tuning knob, and it is simply necessary to turn this knob by hand from one end to the other to produce each required recording.

A very useful review of the swept frequency techniques that were used over the period 1955 to 1966 has been given by Ely.[81]

10.4 Conclusions

Out of the many different configurations that have been described in chronological order in Section 10.3, the one due to D.E. de Jersey[79] gives by far the highest dynamic range. When wideband high-performance single-sideband modulators become more readily available, this measurement technique is likely to grow in popularity.

The techniques that use d.c. amplification[67, 70, 74, 75] are simple but their dynamic range is limited by drift and $1/f$ noise. Better results can be achieved by chopping the microwave source and using a ratiometer that contains tuned amplifiers.[5, 59, 69]

The calibration grid technique[72, 73] removes all worries about going outside the square law region of the detector and it gives a worthwhile improvement in both accuracy and dynamic range. The recently developed techniques for straightening the calibration lines[76, 77] make it much easier to interpret the results given by the calibration grid method.

10.5 References

1 BECK, A.H.W.: 'Velocity-modulated thermionic tubes' (Cambridge University Press, 1948)
2 HUNTON, J.K., and PAPPAS, N.L.: 'The hp-microwave reflectometers', *Hewlett-Packard J.*, 1954, **6**, No. 1–2
3 KINNEAR, J.A.C.: 'An automatic swept-frequency impedance meter', *Br. Commun. Electron.*, 1958, **5**, pp. 359–361
4 SIMS, G.D., and STEPHENSON, I.M.: 'Microwave tubes and semiconductor devices' (Blackie, 1963)
5 LACY, P.D. and WHEELER, D.E.: 'A new 8–12 kMc voltage-tuned sweep oscillator for faster microwave evaluations', *Hewlett-Packard J.*, 1957, **8**, No. 6
6 DUDLEY, R.L.: 'A new series of microwave sweep oscillators with flexible modulation and levelling', *Hewlett-Packard J.*, 1963, **15**, No. 4
7 LACY, P.D., and WHEELER, D.E.: 'New broadband microwave power amplifiers using helix-coupled TWT's', *Hewlett-Packard J.*, 1954, **6**, No. 3–4
8 LACY, P.D. and MATHERS, G.W.C.: 'New TWT amplifiers with provision for simulating special microwave signals', *Hewlett-Packard J.*, 1956, **7**, No. 5
9 CUMMING, R.C.: 'The serrodyne frequency translator', *Proc. IRE*, 1957, **45**, pp. 175–186
10 CONVERT, G. and YEOU, T.: 'Backward wave oscillators', *in* BENSON, F.A. (Ed.): 'Millimetre and submillimetre waves' (Iliffe, 1969)
11 KAKIHANA, S.: 'Current status and trends in high frequency transistors', *Microwave Syst. News*, 1973, **3**, pp. 46–50
12 BULMAN, P.J., HOBSON, G.S. and TAYLOR, B.C.: 'Transferred electron devices' (Academic Press, 1972)
13 HOBSON, G.S.: 'The Gunn effect' (Clarendon Press, Oxford, 1974)
14 GIBBONS, G.: 'Avalanche diode microwave oscillators' (Clarendon Press, Oxford, 1973)
15 ZIEGER, D.: 'Frequency modulation of a Gunn-effect oscillator by magnetic tuning', *Electron. Lett.*, 1967, **3**, (7), 324–325
16 KURU, I.: 'Frequency modulation of the Gunn oscillator', *Proc. IEEE*, 1965, **53**, pp. 1642–1643
17 WARNER, F.L. and HERMAN, P.: 'Miniature X band Gunn oscillator with a dielectric-tuning system', *Electron. Lett.*, 1966, **2**, (12), pp. 467–468
18 WARNER, F.L. and HERMAN, P.: 'Miniature X band Gunn oscillators', Proceedings of the first Cornell University conference on high frequency generation and amplification, August 1967, pp. 206–217
19 LEE, B.K. and HODGART, M.S.: 'Microwave Gunn oscillator tuned electronically over 1 GHz, *Electron. Lett.*, 1968, **4**, (12), pp. 240–242
20 BREHM, G.E. and MAO, S.: 'Varactor tuned integrated Gunn oscillator', *IEEE J. Solid State Circuits*, 1968, **SC-3**, pp. 217–220
21 WILSON, K.: 'Gunn effect devices and their applications', *Mullard Tech. Commun.*, 1969, No. 100, pp. 286–293
22 SMITH, R.B. and CRANE, P.W.: 'Varactor-tuned Gunn-effect oscillator', *Electron. Lett.*, 1970, **6**, (5), pp. 139–140
23 CAWSEY, D.: 'Wide-range tuning of solid state microwave oscillators', *IEEE J. Solid-State Circuits*, 1970, **SC-5**, pp. 82–84

24 CAWSEY, D.: 'Varactor-tuned Gunn-effect oscillators', *Electron. Lett.*, 1970, **6**, pp. 246

25 LARGE, D.: 'Octave band varactor-tuned Gunn diode sources', *Microwave J.*, 1970, **13**, pp. 49–51

26 AITCHISON, C.S., and NEWTON, B.H.: 'Varactor-tuned X band Gunn oscillator using lumped thin-film circuits', *Electron. Lett.*, 1971, **7**, (4), pp. 93–94

27 DOWNING, B.J., and MYERS, F.A.: 'Broadband (1·95 GHz) varactor-tuned X band Gunn oscillator', *Electron. Lett.*, 1971, **7**, (14), pp. 407–409

28 HAMILTON, C.H.: 'Electronically tuned sources for microwave sweepers', *Marconi Instrum.*, 1971, **13**, pp. 35–37

29 DAVIES, R. and NEWTON, B.H.: 'Design trends for Gunn oscillators', *Electron. Equip. News*, 1971, **13**, pp. 18–23

30 BRAVMAN, J., and FREY, J.: 'High performance varactor tuned Gunn oscillators'. Proceedings of the third Cornell University conference on high frequency generation and amplification, 1971, pp. 335–339

31 MESL: 'Wideband varactor-tuned Gunn oscillator', *Microwave J.*, 1972, **15**, pp. 68B-68D

32 JOSHI, J.S.: 'Wide-band varactor-tuned X-band Gunn oscillators in full-height waveguide cavity', *IEEE Trans.*, 1973, **MTT-21**, pp. 137–139

33 DOWNING, B.J., and MYERS, F.A.: 'Q band (38 GHz) varactor-tuned Gunn oscillators', *Electron. Lett.*, 1973, **9**, (11), pp. 244–245

34 BUSHNELL, T.R. and ISAACS, A.T.: 'Wideband varactor-tuned solid-state sources to 20 GHz', *Microwave J.*, 1973, **16**, pp. 45–48

35 AITCHISON, C.S., and DAVIES, R.: 'A varactor-tuned Q-band Gunn oscillator', *Int. J. Electron.*, 1973, **35**, pp. 105–108

36 BULLIMORE, E.D., DOWNING, B.J. and MYERS, F.A.: 'Electronic tuning and stabilisation of Gunn effect oscillators'. European Microwave Conference, A3, 4/4, Sept. 1973

37 JOSHI, J.S. and CORNICK, J.A.F.: 'Some general observations on the tuning characteristics of electro-mechanically tuned Gunn oscillators', *IEEE Trans.*, 1973, **MTT-21**, pp. 582–586

38 GOUGH, R.A. and NEWTON, B.H.: 'An integrated wide-band varactor-tuned Gunn oscillator', *IEEE Trans*, 1973, **ED-20**, pp. 863–865

39 AITCHISON, C.S.: 'Method of improving tuning range obtained from a varactor-tuned Gunn oscillator', *Electron. Lett.*, 1974, **10**, (7), pp. 94–95

40 TEMPLIN, A.S. and GUNSHOR, R.L.: 'Analytic Model for varactor-tuned waveguide Gunn oscillators', *IEEE Trans.*, 1974, **MTT-22**, pp. 554–556

41 PAIK, S.F.: 'Q degradation in varactor tuned oscillators', *IEEE Trans.*, 1974, **MTT-22**, pp. 578–579

42 AITCHISON, C.S.: 'Gunn-oscillator electronic tuning range and reactance compensation: an experimental result at X band', *Electron. Lett.*, 1974, **10**, (23), pp. 488–489

43 CHANG, N.S., HAYAMIZU, T. and MATSUO, Y.: 'YIG tuned Gunn effect oscillator', *Proc. IEEE*, 1967, **55**, p. 1621

44 JAMES, D.A.: 'Wide-range electronic tuning of a Gunn diode by an yttrium-iron garnet (y.i.g.) ferrimagnetic resonator', *Electron. Lett.*, 1968, **4**, (21), pp. 451–452

45 OMORI, M.: 'Octave electronic tuning of a CW Gunn diode using a YIG sphere', *Proc. IEEE*, 1969, **57**, p. 97

46 EASSON, R.M.: 'Design and performance of YIG tuned Gunn oscillators' *Microwave J.*, 1971, **14**, pp. 53–58

47 ELLIS, D.J. and GUNN, M.W.: 'YIG tuning of a stripline Gunn oscillator', *IEEE J. Solid State Circuits*, 1972, **SC-7**, pp. 63–65

48 CLARK, R.J. and SWARTZ, D.B.: 'Combine YIG's with bulk-effect diodes', *Microwaves*, 1972, **11**, pp. 46–53

49 LEP: 'Gunn diodes tuned by YIG sphere', *Microwave J.*, 1972, **11**, p. 58H

50 OYAFUSO, R.T. and BROWN, R.E.: 'Think young with agile YIG and varactor tuning', *Microwaves*, 1972, **11**, pp. 40–43

51 'Fundamental YIG-oscillator – 18 to 26·5 GHz', *Microwave J.*, 1974, **17**, p. 26

52 'High-power solid state M/W sweep oscillators', *Microwave J.*, 1974, **17**, p. 18

53 SORGER, G., HERRERO, J. and VERHOEVEN, H.: 'Design of a broadband RF and M/W sweep generator', *Microwave J.*, 1974, **17**, pp. 51–57

54 TORREY, H.C., and WHITMER, C.A.: 'Crystal Rectifiers' (McGraw-Hill, 1948)

55 WARNER, F.L.: 'Detection' *in* BENSON, F.A. (Ed.): 'Millimetre and submillimetre waves' (Iliffe, 1969)

56 SCHROCK, N.B.: 'A new 10 Mc to 12 kMc coaxial crystal detector mount', *Hewlett-Packard J.*, 1955, **6**, No. 6

57 RILEY, R.B.: 'A new coaxial crystal detector with extremely flat frequency response', *Hewlett-Packard J.*, 1963, **15**, No. 3

58 PRICKETT, R. and RENIHAN, L.: 'New waveguide crystal detectors with flat response', *Hewlett-Packard J.*, 1965, **16**, No. 6

59 VIFIAN, H., DAVID, F.K. and FREDERICK, W.L.: 'A voltmeter for the microwave engineer', *Hewlett-Packard J.*, 1972, **24**, No. 3

60 KYHL, R.L.: 'Directional couplers' *in* MONTGOMERY, C.G. (Ed.): 'Technique of microwave measurements' (McGraw-Hill, 1947)

61 BARNETT, E.F., and HUNTON, J.K.: 'A precision directional coupler using multihole coupling', *Hewlett-Packard J.*, **1952, 3**, No. 7–8

62 BARNETT, E.F.: 'More about the hp-precision directional couplers', *Hewlett-Packard J.*, 1953, **4**, No. 5–6

63 HUNTON, J.K., POULTER, H.C. and REIS, C.S.: 'High-directivity coaxial directional couplers and reflectometers', *Hewlett-Packard J.*, 1955, **7**, No. 2

64 PRICKETT, R.: 'New coaxial couplers for reflectometers, detection and monitoring', *Hewlett-Packard J.*, 1965, **16**, No. 6

65 ANDERSON, R.W. and DENNISON, O.T.: 'An advanced new network analyser for sweep measuring amplitude and phase from 0·1 to 12·4 GHz', Hewlett-Packard J., 1967, **18**, pp. 2–10

66 HUNTON, J.K. and RYALS, A.G.: 'Microwave variable attenuators and modulators using PIN diodes', *IRE Trans.*, 1962, **MTT-10**, pp. 262–273

67 ELY, P.C. and DUDLEY, R.L.: 'Microwave swept-frequency measurements using a feedback-levelled signal source', NEREM Record, 1963, 5, pp. 22–23

68 PAPPAS, N.L.: 'A ratiometer', *IRE Trans.*, 1954, **PGI-3**, pp. 28–34

69 HOIGAARD, J.C.: 'Automated measurements of transmission and reflection – a new instrument based on an old concept', *CPEM Digest*, 1972, pp. 188–190

70 LACY, P.D., and WHEELER, D.E.: 'Permanent record and oscilloscope techniques with the microwave sweep oscillator', Hewlett-Packard J., 1957, **9**, No. 1–2

71 DIX, J.C. and SHERRY, M.: 'A microwave reflectometer display system for 7500 to 11 000 Mc/s', *Electron. Engineering*, 1959, **31**, pp. 24–29

72 HUNTON, J.K. and LORENCE, E.: 'Improved sweep frequency techniques for broadband microwave testing', *Hewlett-Packard J.*, 1960, **12, No. 4**

73 MINCK, J.: 'Improved method cuts errors in swept frequency microwave tests', *Electron. Des.*, 1961, **9**, pp. 136–139

74 'Swept frequency techniques', Hewlett Packard Application Note 65, August 1965

75 BRITTON, R.: 'Microwave ratiometer instrumentation technique gives direct reading display', *Microwave J.*, 1965, **8**, pp. 102–108

76 WEINSCHEL, B.O.: 'Automatic transformation of curved-to-flat calibration lines by a normaliser', *IEEE Trans.*, 1969, **IM-18**, pp. 307–316

77 BRITTON, R.H.: 'Memory improves swept frequency measurements', *Microwave J.*, 1973, **16**, pp. 63–64 and 66

78 WEINSCHEL, B.O.: 'Extending the range of attenuation measurements', *Microwave J.*, 1972, **15**, pp. 23–28 and 40

79 DE JERSEY, D.E.: 'A system for swept frequency measurement of high attenuations', *J. Phys. E.*, 1973, **6**, pp. 964–966

80 DRAPER, D. and HURST, G.J.: 'Microwave signal sources for extended measurement capability', *Marconi Instrum.*, 1973, **14**, pp. 44–48

81 ELY, P.C.: 'Swept-frequency techniques', *Proc. IEEE*, 1967, **55**, pp. 991–1002

Chapter 11

Microwave network analysers

11.1 Introduction

A microwave network analyser is a swept or stepped frequency equipment that measures both the real and imaginary parts of all four *s*-parameters of a 2-port device. Microwave network analysers have completely revolutionised the microwave measurement field in the last few years. The earliest successful equipment of this type was developed by Elliott Brothers (London) Ltd. in the mid-1950s and it has been briefly described by Kinnear.[1] Equipment of this type is described in Section 11.2. In the early 1960s, work on pulse compression and phased array radars focused attention on the phase characteristics of microwave components, and several other firms started to develop swept frequency equipment for measuring phase as well as magnitude. In 1967, two very advanced systems of this type appeared on the commercial market. Using appropriate combinations of swept sources and resolvers, one of these systems[2] was designed to cover the range 100 MHz to 12·4 GHz and the other one was even more ambitious,[3] being designed to give coverage without any gaps all the way from 2 MHz to 40 GHz. Both of these systems, and the earlier Elliott ones, provide a polar display of the magnitude and phase angle of transmission or reflection coefficients against frequency. With a Smith chart graticule over the polar display, the normalised input impedance or the normalised input admittance of a network at any frequency in the band can be immediately seen. In the network analysers introduced in 1967, rectangular displays can also be obtained showing the variation with frequency of transmission gain or loss in dB, transmission phase, return loss in dB or reflection phase.

Since 1967, some further network analysers of this type have become commercially available. Outstanding among these is one described recently by Gorss.[4] This instrument covers the range 400 kHz to 500 MHz, has two measurement channels, a characteristic impedance of either 50 Ω or 75 Ω at the insertion points and a dynamic range of greater than 115 dB. In addition to displaying all of the earlier-mentioned quantities two at a time, it will also give a direct sweep display of group delay against frequency.

All of the swept frequency systems mentioned up to now suffer to some extent from mismatch, tracking and directivity errors, which limit the overall accuracy to about ± 2%. To eliminate these errors and at the same time gain many other advantages, various firms and universities have recently developed computer-controlled microwave network analysers.[5-37] In these analysers, the frequencies of the microwave oscillators are controlled by the computer and are stepped instead of swept across the required bands. At the beginning of each working day, various impedance standards are connected to the equipment and the systematic internal errors are measured at each frequency of interest and stored in the computer. Measurements are then made on the unknown devices under the guidance of a stored computer programme. At each frequency, the computer corrects the raw measurement data from the device under test, using the stored information about the internal systematic errors, and then prints out and displays the corrected results.

In Section 11.2, the various microwave and circuit configurations which are used in network analysers are described; and Section 11.3 gives the full theory of one technique that can be used to correct for internal systematic errors in a computer-controlled network analyser. The other sections in this Chapter deal with: other calibration procedures, errors arising in a microwave network analyser and the evaluation of uncertainties.

11.2 Descriptions of network analysers

In a microwave network analyser, the signal which is transmitted through or reflected from the unknown device is compared in amplitude and phase with a reference signal using:

(*a*) a single-sideband system
(*b*) a twin-channel superheterodyne system, or
(*c*) a modulated sub-carrier system.

A simplified block diagram of the equipment described by Kinnear[1] is shown in Fig. 11.1. This is a single-sideband system. It was developed before backward wave oscillators became commercially available, so the microwave source in it is a motor driven reflex klystron with automatic adjustment of the reflector voltage. The amplitude of the microwave signal is kept constant with an electronically-controlled waveguide attenuator. Part of the output is tapped off by the first directional coupler, changed in frequency by 200 Hz in a motor-driven rotary phase shifter[38] and then fed into the balanced mixer. The rest of the microwave signal is fed through the rotary attenuator and the second directional coupler to the 'unknown'. External waveguide links are connected in position so that either a fraction of the reflected signal from the unknown, or the signal that is transmitted through it, can enter the balanced mixer. Both the amplitude and phase of the 200 Hz

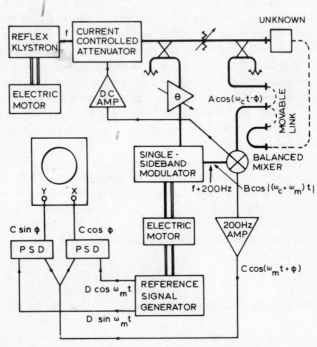

Fig. 11.1 Single-sideband type network analyser described by Kinnear in 1958

signal that emerges from the mixer are precisely related to the amplitude and phase of the microwave signal from the unknown. The required polar display is obtained by following the 200 Hz amplifier with two phase-sensitive detectors that have 200 Hz sine and cosine reference waveforms applied to them. The mathematical expressions included in Fig. 11.1 will help to explain the operation of this system.

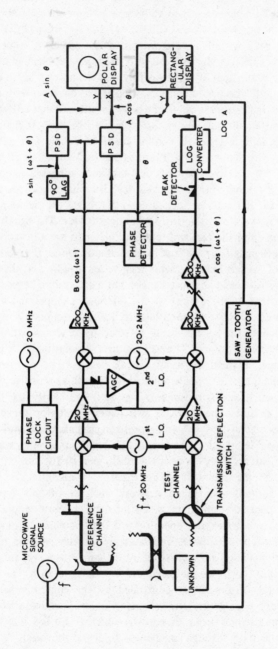

Fig. 11.2 Twin-channel superheterodyne network analyser

Fig. 11.2 contains a simplified block diagram of a non-computer-controlled twin-channel superheterodyne system. The microwave signal source is a backward wave oscillator which is swept over the required range by the sawtooth generator. Part of its output is tapped off by the first coaxial coupler and fed into the reference channel. A similar fraction of its output is tapped off by the second coaxial coupler and applied to the device under test. Depending on the position of the reflection/transmission switch, the power which is either reflected from or transmitted through the unknown is fed into the test channel. The test and reference channels contain identical double superheterodyne receivers, with 1st and 2nd intermediate frequencies of 20 MHz and 200 kHz, respectively. The first local oscillator is phase locked 20 MHz away from the microwave signal source. The phase lock loop has to be designed so that a phase locked condition is always maintained while the BWO is swept over ranges up to an octave wide. The signals entering the mixers from the test and reference channels are kept about 30 dB below the signals from the first local oscillator. Under these conditions, the two i.f. signals have the same relative amplitudes and phases as the microwave reference and test signals. The a.g.c. system keeps the signal level constant at the output of the reference channel and varies the gain of both 20 MHz i.f. amplifiers by the same amount. This arrangement prevents the results being affected by power variations in the microwave signal source. The various displays mentioned in the introduction are obtained by using a peak detector, a log converter,[39] a phase detector, a 90° phase shifter and two phase-sensitive detectors. The mathematical expressions in Fig. 11.2 will clarify the operation of these circuits. The 200 kHz switchable attenuator shown in Fig. 11.2 is used to avoid overloading on strong signals. A network analyser of this type described by Anderson and Dennison[2] gives an attenuation measurement accuracy of about ±0·1 dB per 10 dB, and, in a 6 hour stability test, the total drift did not exceed 0·05 dB and 0·2°.

In a computer-controlled network analyser of this type,[5,6,10] a punched paper-tape program is fed into the computer and a dialogue then occurs between the operator and the computer through a teletypewriter. The microwave oscillator frequencies and the reflection/transmission switch are controlled by the computer in accordance with information typed on the keyboard by the operator. The 200 kHz attenuator is automatically controlled by the computer so that the phase-sensitive detectors always operate under favourable conditions. The output signals from these two detectors are fed into the computer sequentially through an analogue to digital converter.

When the primary microwave source for a network analyser is a backward wave oscillator, the frequency accuracy is about ±0·25%. In the more expensive computer-controlled network analysers, the frequency accuracy is improved several orders of magnitude by deriving the microwave signals from a computer-controlled frequency synthesiser which starts with a quartz oscillator having a stability of a few parts in 10^9 per day. This superior type of source enables precise measurements to be made on high Q cavities and gives a general improvement in repeatability.

By calibrating a network analyser with impedance standards and then using the computer to eliminate the internal systematic errors, a large improvement in the overall accuracy is obtained. To give an example, two 60 dB couplers were measured as carefully as possible against the British national standard and the mean attenuation values obtained were 58·865 dB and 59·564 dB. They were then measured on the HP type 8542A computer-controlled network analyser at RRE and the results obtained were 58·84 dB and 59·56 dB, respectively. The agreements achieved on this occasion were quite remarkable. Several other intercomparisons have been carried out between this network analyser and the British national standards and the results have nearly always been well inside the accuracy specification limits.[12]

In addition to controlling and correcting the measurements, the computer can also be used to do mathematical operations on the output data, typical examples being:

(*a*) conversion of *s* parameters to *h, y* or *z* parameters
(*b*) calculation of the gains obtainable from microwave transistors whose *s* parameters have been measured
(*c*) conversion of reflection coefficients to standing wave ratios
(*d*) calculation of group delay from adjacent points in the phase/ frequency characteristic.

In the computer-operated measurement system described by Davis *et al.*[19] circuit drift errors are eliminated by switching back and forth at a 30 Hz rate between the unknown and a stable microwave reference path. A simplified block diagram of this system is shown in Fig. 11.3. The loss standard contains relay switched precision attenuators which cover the range 0 to 60 dB in 0·01 dB steps. Switches SW3 and SW4 are driven in synchronism with switches SW1 and SW2. The detected 27·7 kHz signal is fed, through an analogue-to-digital converter, into the computer which automatically adjusts the loss standard until the 30 Hz component in the 2nd detector output is almost zero. Thus, this arrangement is basically a driftless automatic i.f. substitution system. The microwave switching is performed with a $17\frac{1}{2}''$ diameter rotating vane. The part of the system which is enclosed by the dotted rectangle in Fig. 11.3 is shown in much greater detail in Fig. 11.4. The mechanical method of switching was adopted as it was found to be more stable and more reproducible than any electronic method of switching. This computer-controlled network analyser gives an amazingly good performance. Accuracies of transmission parameters for small loss are ± 0.002 dB, $\pm 0.01°$, and ± 0.1 nS. This equipment is reported to be 10 times slower than the one described by Hackborn.[5] However, microwave measurements can be carried out on it 300 times faster than they could be using point-by-point techniques.

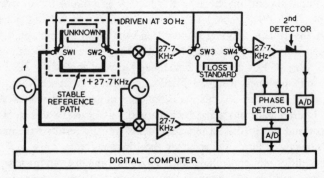

Fig. 11.3 Computer-operated twin-channel superheterodyne network analyser

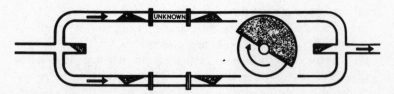

Fig. 11.4 Microwave switching system used by Davis, Hempstead, Leed and Ray

A simplified block diagram of a modulated sub-carrier network analyser is shown in Fig. 11.5. In this system, which stems from a paper by Cohn and Oltman,[40] the microwave signal that is applied to the unknown is amplitude modulated at 200 kHz or thereabouts.

The theory underlying this type of network analyser has already been given in Chapter 8.. The mathematical expressions in Fig. 11.5 show how the required output signals are obtained.

A modulated sub-carrier network analyser has the great advantage of requiring only one microwave source and there is thus no need for the elaborate phase lock loop which is an essential part of a twin-channel superheterodyne system. A computer-programmed modulated

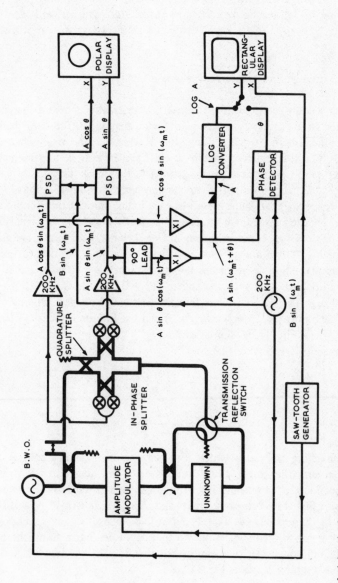

Fig. **11.5** Modulated sub-carrier network analyser

sub-carrier type of network analyser has been developed for use with a time-sharing computer terminal.[14] This approach eliminates the high cost of a special control computer. Shurmer[18] has obtained a similar saving by linking his twin-channel superheterodyne network analyser to a GEC type 90/2 digital computer at the University of Warwick.

11.3 Elimination of internal systematic errors

This section describes how the internal systematic errors can be eliminated in a computer-controlled network analyser of the twin-channel superheterodyne type. Firstly, let us consider the case when the reflection/transmission switch is set for reflection measurements. Then, for the purpose of this analysis, the block diagram given in Fig. 11.2 can be simplified to the configuration shown in Fig. 11.6a.

Assuming that there are no leakage paths and also that the unused port of each directional coupler is perfectly matched, the signal flow graph for Fig. 11.6a is shown in Fig. 11.6b. This signal flow graph contains 24 unknowns and is difficult to analyse, so drastic simplifications will now be made. Let us assume that perfect matches are seen looking back into the source ($\Gamma_g = 0$) and downwards into o_{11} and p_{11}. Also, let us assume that the first coupler is perfectly matched at each port ($m_{11} = m_{22} = m_{33} = 0$) and has infinite directivity ($m_{32} = 0$). Then Fig. 11.6b simplifies to Fig. 11.6c. From the last part of Appendix 2, it follows that

$$t = \frac{o_{21}q_2}{1 - o_{22}q_1} \tag{11.1}$$

and

$$u = \frac{p_{21}r_2}{1 - p_{22}r_1} \tag{11.2}$$

Eliminating the paths which lead only to the matched source, Fig. 11.6c simplifies to Fig. 11.6d. Using the non-touching loop rule given in Appendix 2, the outputs from the reference and test channels are seen from Fig. 11.6d to be $Em_{31}t$ and $Em_{21}\{n_{31}u + (n_{21}\Gamma_L n_{32}u)/(1 - n_{22}\Gamma_L)\}$, respectively. Fig. 11.6d still contains too many unknowns but it can be redrawn as shown in Fig. 11.6e. Again using the non-touching loop rule, it is easily seen that Fig. 11.6e gives exactly the same output voltages as Fig. 11.6d.

Let $Em_{31}t = \xi$, $\dfrac{m_{21}n_{31}u}{m_{31}t} = e_0$ and $\dfrac{m_{21}n_{21}n_{32}u}{m_{31}t} = e_1$ $\left.\begin{array}{c} (11.3) \\ (11.4) \\ (11.5) \end{array}\right\}$

Then, we finish up with the signal flow graph shown in Fig. 11.6*f* which contains only three unknowns.

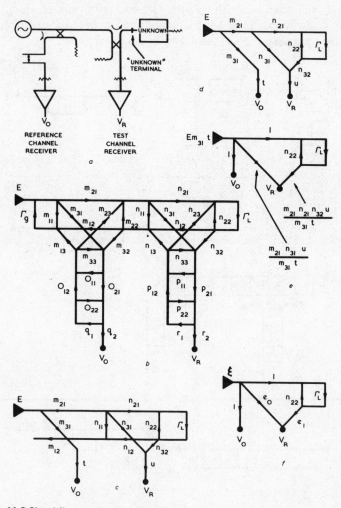

Fig. 11.6 Signal flow graphs for reflection

When the reflection/transmission switch is set for transmission measurements, we can use the simplified block diagram shown in Fig. 11.7*a*. The signal flow graph for Fig. 11.7*a* is shown in Fig. 11.7*b*, where s_{11}, s_{21}, s_{12} and s_{22} are the four *s*-parameters describing the unknown. Proceeding now in a similar way to before, Fig. 11.7*b* can be progressively simplified to Fig. 11.7*f* where

$$x = \frac{v_{21}w_2}{1 - v_{22}w_1} \quad \text{and} \quad e_2 = \frac{m_{21}n_{21}x}{m_{31}t} \quad (11.6), (11.7)$$

Thus, altogether, we finish up with 4 internal unknowns, e_0, e_1, e_2 and n_{22}, whose values must be determined at each frequency of interest before the s-parameters of the unknown can be calculated.

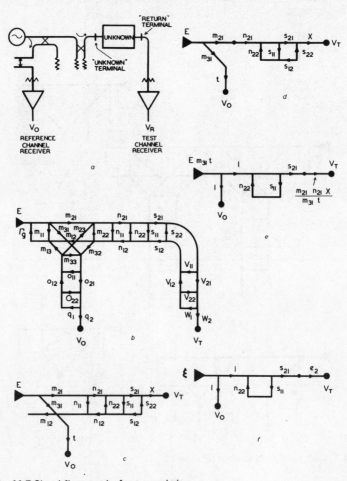

Fig. 11.7 Signal flow graphs for transmission

The values of these four internal unknowns are found by carrying out a 4-stage calibration process. From Fig. 11.6f we see that

$$\frac{V_R}{V_0} = e_0 + \frac{\Gamma_L e_1}{1 - n_{22}\Gamma_L} \quad (11.8)$$

and from Fig. 11.7*f*, it follows that

$$\frac{V_T}{V_0} = \frac{s_{21}e_2}{1 - n_{22}s_{11}} \tag{11.9}$$

The four stages of the calibration process are as follows:

(*a*) *Reflection with matched termination*
First of all, Γ_L is made equal to zero by placing a high class matched termination on the 'unknown' terminal. Then, from eqn. 11.8, we get

$$\frac{V_R}{V_0} = M_1 = e_0 \tag{11.10}$$

(*b*) *Reflection with direct short*
Secondly, Γ_L is made equal to -1 by placing a direct short circuit on the 'unknown' terminal. From eqn. 11.8 we now get

$$\frac{V_R}{V_0} = M_2 = e_0 - \frac{e_1}{1 + n_{22}} \tag{11.11}$$

(*c*) *Reflection with offset short*
The next operation involves placing an offset short of length l on the 'unknown' terminal. We then have

$$\Gamma_L = -e^{-j2\beta l} = \Gamma_s \tag{11.12}$$

Substituting this value into eqn. 11.8, we get

$$\frac{V_R}{V_0} = M_3 = e_0 + \frac{\Gamma_s e_1}{1 - n_{22}\Gamma_s} \tag{11.13}$$

Knowing the values of l and the phase constant β, the value of Γ_s is easily calculated by the computer at each frequency of interest.

(*d*) *Transmission with through connection*
Finally, the 'unknown' terminal is connected directly to the 'return' terminal and a transmission measurement is made. In this case, $s_{11} = s_{22} = 0$ and $s_{12} = s_{21} = 1$, so, from eqn. 11.9 it follows that

$$\frac{V_T}{V_0} = M_4 = e_2 \tag{11.14}$$

From eqns. 11.10, 11.11, 11.13 and 11.14 we get:

$$e_0 = M_1 \tag{11.15}$$

$$e_1 = \frac{(M_1 - M_2)(M_3 - M_1)(1 + \Gamma_s)}{\Gamma_s(M_3 - M_2)} \qquad (11.16)$$

$$e_2 = M_4 \qquad (11.17)$$

$$n_{22} = \frac{(M_3 - M_1) + \Gamma_s(M_2 - M_1)}{\Gamma_s(M_3 - M_2)} \qquad (11.18)$$

When an unknown 2-port is connected to the 'unknown' terminal and a perfectly matched termination is placed behind it, $\Gamma_L = s_{11}$. Hence, from eqn. 11.8 we get

$$\frac{V_R}{V_0} = M_{rf} = e_0 + \frac{s_{11}e_1}{1 - n_{22}s_{11}} \qquad (11.19)$$

Rearranging, we get

$$s_{11} = \frac{M_{rf} - e_0}{e_1 + n_{22}(M_{rf} - e_0)} \qquad (11.20)$$

When a transmission measurement is made with the unknown 2-port in position,

$$\frac{V_T}{V_0} = M_{tf} = \frac{s_{21}e_2}{1 - n_{22}s_{11}} \qquad (11.21)$$

Thus

$$s_{21} = \frac{M_{tf}(1 - n_{22}s_{11})}{e_2} \qquad (11.22)$$

The other two s parameters, s_{22} and s_{12}, can now be found in a similar manner by reversing the unknown and repeating both the reflection and transmission measurements.

All of the parameters used in this section are complex quantities having both real and imaginary parts. This fact makes the computer's task much more complicated than it may appear to be at first sight.

11.4 Other calibration procedures

The procedure described in Section 11.3 is only one of many that can be used. In the procedure described by Hand,[12] a leakage path is included and it is not assumed that there is a perfect match looking into the receiver. Thus, there are 6 unknowns in the final signal flow graphs and a 6-stage calibration procedure has to be carried out. Instead of using a fixed matched termination, a sliding one is employed and this is measured at four different positions. The computer then determines the centre of the circle that passes through the four measured values.

This operation eliminates errors that would otherwise be caused by the small residual reflection of the sliding termination. After completing this operation, the other five stages of the calibration procedure are: reflection with a direct short, reflection with an offset short, reflection and transmission with a through connection and, finally, transmission without a through connection (both measurement ports terminated). This six stage calibration procedure results in quite complicated equations which are solved by iterative processes in the computer-corrected microwave network analysers that are made by Hewlett Packard. However, Kruppa and Sodomsky,[16] Davis, Doshi and Nagenthiram,[27] Rehnmark[28] and Gelnovatch[37] have shown that explicit solutions can be found for these equations. By using explicit formulae instead of iterative methods, the computational efficiency can be improved.

Instead of performing the calibration with a matched termination, a direct short and an offset short, it can be done with:

(a) a direct short and two offset shorts
(b) a direct short, an offset short and an open circuit
(c) a matched termination, a direct short and a standard mismatch
(d) a direct short, an offset short and a sliding termination whose reflection coefficient need not be known
(e) a direct short and two sliding terminations − the first with a high reflection coefficient and the second with a low one, neither of which need be known
(f) a direct short, a through line and a delay line.

Method (a) was proposed by Silva and McPhun[20] and it is particularly suitable for microstrip work and also for tasks where space restrictions rule out the use of a precision matched sliding termination. Care must be exercised when choosing the offset lengths as certain values can cause overloading in the computer.

Method (b) was devised by Shurmer[22] for carrying out measurements on microwave transistors mounted in a microstrip jig. The open circuit is produced by simply leaving the 'transistor well' empty. A discontinuity capacitance is then present but this can be taken into account.[41]

Gould and Rhodes[26] described method (c). Their standard mismatch is a commercially-available item with a v.s.w.r. of $1 \cdot 5 \pm 5\%$, over the band 1 to 4 GHz. This calibration technique does not require precise measurements of the signal frequency as the three standards are frequency insensitive and they are all used at the same plane.

Methods (d) and (e) were devised by Kasa.[31] The full theory underlying these two techniques has been published.[33] The calibration constants are obtained by direct calculation. Kasa[33] neglected the line attenuation in the sliding terminations but Engen[34] has included it in a recent treatment.

Franzen and Speciale[36] described method (*f*). The standards required for this method do not contain any critical mechanical parts and the equations that are associated with it have an exact explicit solution.

Woods[35] has pointed out that the calibration procedure described by Hand[12] overlooks the small impedance changes that occur when the reflection/transmission switches are changed over. To avoid this shortcoming, he has suggested that these switches should be eliminated by using three receiving channels. He has carried out a completely rigorous analysis of this proposed new scheme. By adopting it, the random errors caused by non-repeatability of the switches would be eliminated and two 10 dB pads would become unnecessary, so the dynamic range would be improved. Three single-port standards would be needed to do the calibration.

11.5 Errors arising in a microwave network analyser

Several sources of error have already been mentioned and a complete list is given below:

(*a*) imperfect matching at the insertion point
(*b*) non-infinite directivity in the couplers
(*c*) departures from perfect tracking in the two channels
(*d*) leakage
(*e*) noise
(*f*) imperfections in the calibration standards (these are caused by incorrect diameters for the inner and outer conductors, an error in the length of the offset short, contact resistances between the inner and outer conductors, eccentricity in the coaxial lines, line losses and discontinuity capacitances)
(*g*) imperfections in the instrumentation (these include frequency errors, power variations in the microwave signals, non-repeatability in the microwave switches, non-linearity in the mixers, gain and phase drifts in the i.f. amplifiers, d.c. drift in the a.g.c. loop, errors in the 200 kHz attenuator, quadrature errors in the phase sensitive detectors, errors in the analogue to digital converter etc.)
(*h*) impedance changes that occur when the reflection/transmission

switches are changed over

(*i*) connector uncertainties.

In a computer-corrected microwave network analyser, we have seen that errors (*a*), (*b*) and (*c*) are eliminated. If leakage paths are included in the signal flow graphs, error (*d*) is also eliminated. Error (*h*) was discussed in Section 11.4 and error (*i*) can be minimised by using precision coaxial connectors such as the GR900 or APC-7 (see Chapter 5). Drift errors can be eliminated by using the technique described by Davis *et al.*[19] Thus, if all possible steps are taken, we are still left with errors (*e*) and (*f*) and various instrumental errors.

Adam[6] has given equations for the magnitude and phase uncertainties that occur when making reflection measurements on an HP 8542A network analyser with phase locked sources. These are as follows:

$$\text{magnitude uncertainty} = \pm(0{\cdot}0015 + 0{\cdot}005\,|s_{11}| + 0{\cdot}003\,|s_{11}|^2)$$

$$(11.23)$$

$$\text{angle uncertainty} = \pm\left\{0{\cdot}25^{\circ} + \tan^{-1}\frac{0{\cdot}0015}{|s_{11}|} + 4\tan^{-1}(0{\cdot}005\,|s_{11}|)\right\}$$

$$(11.24)$$

Hand[12] has made a detailed study of the errors caused by (*e*), (*f*) and (*g*) in the HP8542A. He used a random number generator to assign random phase angles between 0 and 2π and random magnitudes (between specification limits) to all of the variables that contribute to these errors. He fed these quantities into a computer program that yields a set of calculated *s*-parameters which are compared with the input values to determine the errors. This process is repeated 100 times, with new random magnitudes and phase angles each time. The ninth worst error is then printed out and this gives the measurement uncertainty with a 96% confidence level. For the results thus obtained, Hand's paper[12] should be consulted.

Ridella[23, 30] has derived equations for the calibration repeatability, the measurement repeatability, the residual mismatch etc., and he has also reported the results that were obtained when various devices were measured on four different automatic network analysers.

11.6 Evaluation of uncertainties

A computer-corrected microwave network analyser is an extremely complex instrument and, occasionally, users wonder if a small hidden fault is producing any errors in the thousands of results that it prints out each hour. A way of restoring faith is to measure on it each week a precisely calculable 2-port standard and see if there are any significant differences between the measured and theoretical values. Three different devices that have been considered for this role are:

(a) a half-round inductive obstacle in a short length of rectangular waveguide[42]
(b) a capacitive rod in a short length of rectangular waveguide[43]
(c) a section of coaxial line or waveguide, a quarter wavelength long at midband, with a characteristic impedance different from the normal value.[44-46]

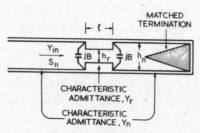

Fig. 11.8 Reduced height waveguide reflection coefficient standard

Item (c) is the one that has been most widely used. Only the reflection characteristics of the waveguide version of item (c) will be discussed here. Consider the configuration shown in Fig. 11.8 where h_n denotes the normal waveguide height and h_r and l denote the height and length of the 2-port standard. The discontinuity capacitances must be taken into account and their susceptances are denoted by B. Then, neglecting losses, the admittance looking into the 2-port standard is seen to be:

$$Y_{in} = jB + Y_r \frac{(Y_n + jB) + jY_r \tan \theta)}{Y_r + j(Y_n + jB) \tan \theta} \qquad (11.25)$$

and

$$s_{11} = \frac{Y_n - Y_{in}}{Y_n + Y_{in}} = - \frac{1 - \alpha^2 - \alpha^2 b^2 + \dfrac{2\alpha b}{\tan \theta}}{1 + \alpha^2 - \alpha^2 b^2 + \dfrac{2\alpha b}{\tan \theta} - j\left(\dfrac{2\alpha}{\tan \theta} - 2\alpha^2 b\right)} \qquad (11.26)$$

where $\qquad \alpha = \dfrac{h_r}{h_n} = \dfrac{Y_n}{Y_r}, \quad b = \dfrac{B}{Y_n} \quad$ and $\quad \theta = \dfrac{2\pi l}{\lambda_g} \left.\begin{array}{r}(11.27)\\(11.28)\\(11.29)\end{array}\right\}$

It is now a straight-forward matter to split eqn. 11.26 into real and imaginary parts or find expressions for $|s_{11}|$ and the reflection phase angle.

Marcuvitz[47] gives the following approximate expression for the normalised discontinuity susceptance:

$$b \approx \frac{2h_n}{\lambda_g} \left\{ \frac{\alpha^2+1}{2\alpha} \log_e \frac{1+\alpha}{1-\alpha} + \log_e \frac{1-\alpha^2}{4\alpha} + \frac{2}{\Lambda} \right\} \qquad (11.30)$$

where

$$\Lambda = \left(\frac{1+\alpha}{1-\alpha}\right)^{2\alpha} \cdot \frac{1+[1-(h_n/\lambda_g)^2]^{1/2}}{1-[1-(h_n/\lambda_g)^2]^{1/2}} - \frac{1+3\alpha^2}{1-\alpha^2} \qquad (11.31)$$

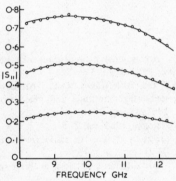

Fig. 11.9 Calculated and measured values of $|s_{11}|$ versus frequency for three 2-port transfer standards
———— theoretical curves
oooo measured values on the RRE ANA

Beatty[45] has reported that measured values for b are 4% higher than the values given by eqn. 11.30. Fig. 11.9 shows calculated and measured values of $|s_{11}|$ versus frequency for three 2-port transfer standards of this type. The theoretical curves were obtained from eqns. 11.26 to 11.31 and the 4% correction factor was included. The measured values were obtained on the HP 8542A automatic network analyser at RRE.

11.7 Conclusions

Modern network analysers provide complete magnitude and phase information about one and two-port networks. This information is provided almost instantaneously over an enormous frequency range, with an accuracy in the region of ± 2%. The recently introduced computer-controlled network analysers are even more exciting instruments.

By the use of very ingenious calibration and computer correction techniques, these automatic network analysers are able to give results which are almost as accurate as those obtained in the best microwave standards laboratories. The speed with which measurements can be made by a computer-controlled network analyser is almost unbelievable. Tasks which would take several days to perform using point-by-point techniques can be completed in a few minutes on them. The computer-controlled network analyser has revolutionised the microwave measurement field and is an excellent example of successful automation.

11.8 References

1 KINNEAR, J.A.C.: 'An automatic swept-frequency impedance meter', *Br. Commun. & Electron.*, 1958, 5, pp. 359–361

2 ANDERSON, R.W., and DENNISON, O.T.: 'An advanced new network analyser for sweep-measuring amplitude and phase from 0·1 to 12·4 GHz', *Hewlett-Packard J.*, 1967, 18, pp. 2–10

3 Wiltron Company Catalogue, 1968, pp. 2–5

4 GORSS, C.G.: 'A precision system for radio frequency network analysis', *IEEE Trans.*, 1972, IM-21, pp. 538–543

5 HACKBORN, R.A.: 'An automatic network analyser system', *Microwave J.*, 1968, 11, pp. 45–52

6 ADAM, S.F.: 'A new precision automatic microwave measurement system', *IEEE Trans.*, 1968, IM-17, pp. 308–313

7 ADAM, S.F.: 'Microwave theory and applications' (Prentice-Hall, 1969), pp. 445–454

8 SHURMER, H.V.: 'Correction of a Smith-chart display through bilinear transformations', *Electron. Lett.*, 1969, 5, p. 209

9 GELDART, W.J., *et al.*: 'A 50 Hz-250 MHz computer operated transmission measuring set', *BSTJ*, 1969, 48, pp. 1339–1381

10 RYTTING, D.K., and SANDERS, S.N.: 'A system for automatic network analysis', *Hewlett-Packard J.*, 1970, 21, pp. 2–10

11 RAY, W.A., and WILLIAMS, W.W.: 'Software for the automatic network analyser', *Hewlett-Packard J.*, 1970, 21, pp. 11–15

12 HAND, B.P.: 'Developing accuracy specifications for automatic network analyser systems', *Hewlett-Packard J.*, 1970, 21, pp. 16–19

13 HUMPHRIES, B.: 'Applications of the automatic network analyser', *Hewlett-Packard J.*, 1970, 21, pp. 20–24

14 Wiltron Company Catalogue, 1970, pp. 20–23

15 SHURMER, H.V.: 'New method of calibrating a network analyser', *Electron. Lett.*, 1970, 6, pp. 733–734

16 KRUPPA, W., and SODOMSKY, K.F.: 'An explicit solution for the scattering parameters of a linear two-port measured with an imperfect test set', *IEEE Trans.*, 1971, MTT-19, pp. 122–123

17 SHURMER, H.V.: 'On-line computer correction of microwave measurements on semiconductor devices', IEE Colloquium Digest 1971/13, 1971, pp. 4/1–4/6

18 SHURMER, H.V.: 'Low-level programming for the on-line correction of microwave measurements', *Radio & Electron. Eng.*, 1971, **41**, pp. 357–364

19 DAVIS, J.B., *et al.*, '3700–4200 MHz computer operated measurement system for loss, phase, delay and reflection', *IEEE Trans.*, 1972, **IM-21**, pp. 24–37

20 daSILVA, E.F., and McPHUN, M.K.: 'Calibration of microwave network analyser for computer-corrected *s*-parameter measurements', *Electron. Lett.*, 1973, **9**, pp. 126–128

21 DAVIS, R.T.: 'Automation upgrades microwave measurements', *Microwaves*, 1973, **12**, pp. 40–44 and 77

22 SHURMER, H.V.: 'Calibration procedure for computer-corrected *s*-parameter characterisation of devices mounted in microstrip', *Electron. Lett.*, 1973, **9**, pp. 323–324

23 RIDELLA, S.: 'Computerised microwave measurements accuracy analysis'. Proceedings of the European Microwave Conference, Paper B34, 1973

24 FITZPATRICK, J.K.: 'A new direction for automatic network analyser software', *Microwave J.*, 1973, **16**, pp. 65–68

25 ARNOLD, S., and WHITELEY, I.: 'Computer controlled measurement of 2- and 3-port microwave devices' *in* 'The use of digital computers, in measurement'. IEE Conf. Publ. 103, 1973, pp. 125–130

26 GOULD, J.W., and RHODES, G.M.: 'Computer correction of a microwave network analyser without accurate frequency measurement', *Electron. Lett.*, 1973, **9**, pp. 494–495

27 DAVIES, O.J., DOSHI, R.B., and NAGENTHIRAM, B.: 'Correction of microwave network analyser measurements of 2-port devices', *Electron. Lett.*, 1973, **9**, pp. 543–544

28 REHNMARK, S.: 'On the calibration process of automatic network analyser systems', *IEEE Trans.*, 1974, **MTT-22**, pp. 457–458

29 BEATTY, R.W.: 'Methods for automatically measuring network parameters', *Microwave J.*, 1974, **17**, pp. 45–49 and 63

30 RIDELLA, S.: 'Computerised reflection measurements', *CPEM Digest*, 1974, pp. 51–53

31 KASA, I.: 'Exact solution of network analyzer calibration and two-port measurements by sliding terminations', *CPEM Digest*, 1974, pp. 90–92

32 LITTLE, W.E.: 'Automated computer controlled measurements', *CPEM Digest*, 1974, pp. 331–332

33 KASA, I.: 'Closed-form mathematical solutions to some network analyser calibration equations', *IEEE Trans.*, 1974, **IM-23**, pp. 399–402

34 ENGEN, G.F.: 'Calibration technique for automated network analyzers with application to adapter evaluation', *IEEE Trans.*, 1974, **MTT-22**, pp. 1255–1260

35 WOODS, D.: 'Rigorous derivation of computer-corrected network-analyser calibration equations', *Electron. Lett.*, 1975, **11**, pp. 403–405

36 FRANZEN, N.R., and SPECIALE, R.A.: 'A new procedure for system calibration and error removal in automated *s*-parameters measurements', Proceedings of the 5th European Microwave Conference, pp. 69–73, 1975

37 GELNOVATCH, V.G.: 'A computer program for the direct calibration of two-port reflectometers for automated microwave measurements', *IEEE Trans.*, 1976, **MTT-24**, pp. 45–47

38 FOX, A.G.: 'An adjustable waveguide phase changer', *Proc. IRE*, 1947, 35, pp. 1489–1498

39 GIBBONS, J.F., and HORN, H.S.: ' A circuit with logarithmic transfer response over 9 decades', *IEEE Trans.* 1964, **CT-11**, pp. 378–384

40 COHN, S.B., and OLTMAN, H.G.: 'A precision microwave phase-measurement system with sweep presentation', *IRE Int. Conven. Rec.*, 1961, 9, Pt. 3, pp. 147–150

41 JAMES, D.S., and TSE, S.H.: 'Microstrip end effects', *Electron. Lett.*, 1972, 8, 46–47

42 KERNS, D.M.: 'Half-round inductive obstacles in rectangular waveguide', *J. Res. NBS*, 1960, **64B**, pp. 113–130

43 LEWIN, L.: 'Standard mismatch: the production of controlled small reflections in waveguides', *J. Res. NBS*, 1968, **72C**, pp. 197–201

44 BEATTY, R.W.: '2-port $\lambda_G/4$ waveguide standard of voltage standing-wave ratio', *Electron. Lett.*, 1973, 9, pp. 24–26

45 BEATTY, R.W.: '2-port standards for evaluating automatic network parameter measurement systems', *CPEM Digest*, 1974, pp. 87–89

46 BEATTY, R.W.: 'Calculated and measured s_{11}, s_{21}, and group delay for simple types of coaxial and rectangular waveguide 2-port standards', *NBS Technical Note 657*, Dec. 1974

47 MARCUVITZ, N.: 'Waveguide handbook' (McGraw-Hill, 1951), pp. 307–309

Chapter 12

Other methods of measuring microwave attenuation

In this chapter, several less-well-known methods of measuring microwave attenuation are described.

12.1 Laverick's three channel technique

Fig. 12.1 shows the equipment needed for this technique.[1] The measurement procedure is as follows:

(a) With SW1 closed, SW2 open and the unknown set at maximum attenuation, A_1 and θ_1 are adjusted until the output is zero
(b) With SW1 open and SW2 closed, A_2 and θ_2 are adjusted until the output is again zero
(c) With both switches closed, the unknown and θ_1 are adjusted until the output is yet again zero.

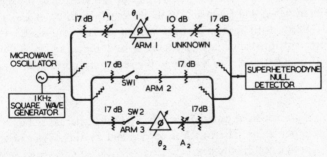

Fig. 12.1 Laverick's self-calibrating attenuation measurement system

Steps b and c are then repeated over and over again. The voltage ratios corresponding to successive movements of the unknown attenuator are 2/1, 3/2, 4/3, 5/4 etc.

Table 12.1 summarises the complete process. This method requires a large number of waveguide components and its suffers from several sources of error which have been analysed by Laverick.[1] However, with care, it will give an accuracy of about ± 0·02 dB up to 20 dB.

Table 12.1

Number of operation	Output voltage from arm 1 before adjustment of unknown	Output voltage from arm 2 with SW1 closed	Output voltage from arm 3 after step b	Combined output voltage from arms 2 & 3 during step c	Output voltage from arm 1 after adjustment of unknown	Attenuation change in unknown dB
1	+ e	− e	− e	− 2e	+ 2e	6·021
2	+ 2e	− e	− 2e	− 3e	+ 3e	3·522
3	+ 3e	− e	− 3e	− 4e	+ 4e	2·499
4	+ 4e	− e	− 4e	− 5e	+ 5e	1·938
5	+ 5e	− e	− 5e	− 6e	+ 6e	1·584
6	+ 6e	− e	− 6e	− 7e	+ 7e	1·339

12.2 Peck's self-calibration technique

Peck[2] has described four different circuits that can be used to calibrate a variable attenuator without the use of an attenuation standard. One of his circuits is shown in Fig. 12.2. A_1 is a fixed attenutator, A_2 is an uncalibrated variable attenuator and A_u is the unknown, which will be assumed to be a microwave piston attenuator in the following explanation. The two phase shifters, ϕ_1 and ϕ_2 are set so that the signals from branches A and C are always in phase with each other and 180° out of phase with the signal from branch B. The four steps of the measurement procedure are given below and they can be followed easily if frequent reference is made to the vector diagrams shown in Fig. 12.3.

Step 1 A_u is set to a reading l_1 which gives the highest required attenuation value. The switch is opened and ϕ_2 and A_2 are adjusted until a null is obtained.

Step 2 The switch is closed adding a voltage V_A to V_{C1} and the unknown is then changed to a lower setting l_2 which restores the null.

Step 3 The switch is opened and A_2 is adjusted until a null is achieved.

Step 4 The switch is closed again, adding a voltage V_A to V_{C2}, and the unknown is moved to an even lower setting l_3, which produces a null for the fourth time.

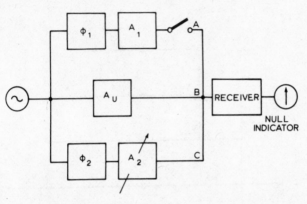

Fig. 12.2 Equipment needed for Peck's self-calibration technique

Let the piston attenuator provide an attenuation change of A_p decibels when the separation is increased by unit length. Then, it follows from Fig. 12.3 that:

$$A_p(l_1 - l_3) = 20 \log_{10} \frac{V_{B3}}{V_{B1}} = 20 \log_{10} \frac{V_{B1} + 2V_A}{V_{B1}} = 20 \log_{10} \frac{k+2}{k}$$

(12.1)

and

$$A_p(l_2 - l_3) = 20 \log_{10} \frac{V_{B3}}{V_{B2}} = 20 \log_{10} \frac{V_{B1} + 2V_A}{V_{B1} + V_A} = 20 \log_{10} \frac{k+2}{k+1}$$

(12.2)

where $V_{B1}/V_A = k$. Then, dividing eqn. 12.2 by eqn. 12.1, we get:

$$\frac{l_2 - l_3}{l_1 - l_3} = \frac{\log_{10} \dfrac{k+2}{k+1}}{\log_{10} \dfrac{k+2}{k}}$$

(12.3)

The value of k can easily be found with an iterative computer program and, when it is known, A_p can be determined from either eqn. 12.1 or eqn. 12.2.

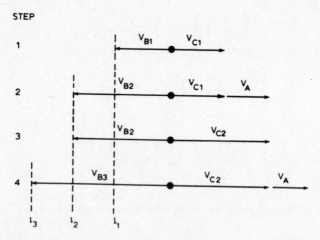

Fig. 12.3 Vector diagrams for the four stages of Peck's self-calibration technique

Peck used this method to measure the attenuation coefficient of a 30 MHz piston attenuator and he obtained a value of 10·003 dB/inch with a standard deviation of 0·003 dB/inch. The value of the attenuation coefficient computed from theoretical considerations was 10·000 ± 0·002 dB/inch.

12.3 Two-channel null method

In 1964, Iwase[3] described a two-channel null technique which is very suitable for measuring small changes in attenuation that are not accompanied by significant changes in phase. A simplified block diagram of the equipment needed for this technique is shown in Fig. 12.4.

The measurement procedure is as follows:

(a) with the switch open and the unknown set at zero, attenuator A is set to a reading A_1 which gives a convenient output on the indicator

(b) the switch is now closed, A is set to a low value to give high sensitivity and then, A_R and θ_R are adjusted until a null is obtained

(c) the unknown is now changed by a small amount ΔA_u, and A is adjusted to a setting A_2 which gives the same output as before (*d*) the change in attenuation through the unknown is calculated from the following equation:

$$\Delta A_u = 20 \log_{10} \frac{1}{1 - 10^{-(A_1 - A_2)/20}} \quad (12.4)$$

This equation can be derived as follows. At the completion of stage (*a*), let the signal amplitudes at the input and output of attenuator A be e_1 and e_{out}, respectively. Then, with perfect matching,

$$A_1 = 20 \log_{10} \frac{e_1}{e_{out}} \quad (12.5)$$

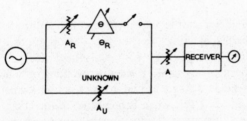

Fig. 12.4 Two-channel null technique due to Iwase

At the end of stage (*b*), e_1 will be completely cancelled out by an equal and opposite signal from the upper channel. Let the change ΔA_u reduce the lower channel signal from e_1 to e_2. Then

$$\Delta A_u = 20 \log_{10} \frac{e_1}{e_2} \quad (12.6)$$

and

$$A_2 = 20 \log_{10} \frac{e_1 - e_2}{e_{out}} \quad (12.7)$$

Simple manipulation of eqns. 12.5, 12.6 and 12.7 rapidly leads to eqn. 12.4.

With this technique, very small attenuation changes through the unknown are associated with very large differences between A_1 and A_2 (see Table 12.2). A variation of this technique, which makes use of two off-null conditions, has been described by Nemoto, Beatty and Fentress.[4] Good agreement has been obtained[5] between this off-null technique and the power ratio method.[6] Nemoto[7] has used this off-null technique to determine the properties of waveguide joints.

Table 12.2

$A_1 - A_2$	ΔA_u
dB	dB
20	0·9151
30	0·2791
40	0·0873
50	0·0275
60	0·0087

12.4 Single-oscillator shuttle pulse method

The single-oscillator shuttle pulse method of measuring attenuation was proposed initially by Ring[8] and the first description of it in the published literature was given by King and Mandeville[9] in 1961. Since then it has been widely used by scientists working on millimetre wave low-loss waveguide communication systems at the Marconi Company in England, the Bell Telephone Laboratories in the USA, The Nippon Telegraph and Telephone Corporation in Japan and the Post Office Research Department in England. A block diagram of a typical shuttle pulse system is shown in Fig. 12.5. For most of the time, the backward wave oscillator has constant voltages applied to its electrodes and it provides a local oscillator signal at a constant frequency f_0. However, flat-topped pulses with a duration somewhere in the range 10 ns to $1 \mu S$ and a p.r.f. in the region of 10 kHz are applied to the helix of this backward wave oscillator and the amplitude of these pulses is chosen so that the frequency is shifted from f_0 to f_1 while they are present. The ferrite circulator directs the output from the backward wave oscillator, through a rectangular to circular transformer, to the length of waveguide under test, which is provided with a coupling mesh at the input end and a short circuit at the far end. The coupling mesh is designed to transmit about 1% of the incident power. The pulses at frequency f_1 bounce back and forth many times between the coupling mesh and the short circuit. After each round trip a fraction of this power passes back through the coupling mesh and enters the mixer. Here, it is mixed with the strong signal that is reflected by the front face of the coupling mesh. The two-way delay through the waveguide under test is arranged to be longer than the pulse duration, so this strong signal is at a frequency f_0 when the train of pulses at frequency f_1 arrive. Mixer output pulses at an intermediate frequency

$f_1 - f_0$ are therefore produced and these are amplified, detected and displayed on an oscilloscope.

Let A_1 denote the attenuation constant in dB per unit length of the waveguide under test, let L be the distance between the coupling mesh and the short circuit and let A_m and A_s denote the losses in dB that occur at each reflection from the mesh and short circuit, respectively. Then, the total loss in dB that occurs in each successive two-way trip through the test section is given by

$$A_t = 2A_1 L + A_m + A_s \qquad (12.8)$$

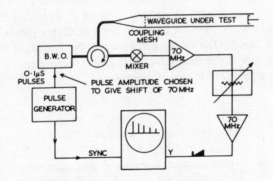

Fig. 12.5 Shuttle pulse method of measuring attenuation

The value of A_t can be determined from any adjacent pair of pulses on the oscilloscope. Additional measurements are needed to find the values of A_m, A_s and L, and then A_1 can be obtained from eqn. 12.8.

An output power of 1 to 5 mW from the backward wave oscillator is entirely adequate. A one-way loss of 20 dB through the coupling mesh keeps the strongest echo 40 dB below the local oscillator signal, and, with this ratio, excellent mixer linearity is obtained. The coupling mesh usually takes the form of a flat transverse metal plate containing many small uniformly spaced holes. With short test sections, as many as 100 echoes can be seen. Range gating techniques can be used to select the various echoes one by one and sophisticated digital circuits can be used to measure the echo amplitudes very accurately. Alternatively, r.f. substitution or i.f. substitution techniques can be used to bring successive gated echoes to the same output level.

Table 12.3 summarises the main characteristics of nine different shuttle pulse equipments that have been described in the literature.

The systems described by Moorthy[16] and Bomer[17] are controlled by minicomputers. In addition to using the shuttle pulse technique, Shimba[13,15] has used a single reflection pulse method. In this method, a coupling grid is not used.

12.5 Determination of attenuation from Q measurements

Very accurate measurements of the attenuation per unit length in a uniform piece of waveguide can be obtained by using the arrangement shown in Fig. 12.6. Two non-contacting sliding short circuits are placed inside the waveguide so that a cavity is formed. The coaxial lines and adjustable coupling loops are used to couple power into and out of the cavity. The distance l between the two short circuits is adjusted to several different values that give resonance and, at each of these settings, the Q of the cavity is measured.

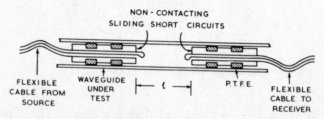

Fig. 12.6 Q method of measuring attenuation

By slight rearrangement of the cavity equations given by Barlow and Cullen,[18] it is easily shown that the Q is given by

$$Q = \frac{2\pi l \lambda_g}{\lambda^2 (2 - |\Gamma_1| - |\Gamma_2| + 2\alpha_1 l)} \quad (12.9)$$

where α_1 is the attenuation per unit length in Nepers, λ_g is the guide wavelength, λ is the free space wavelength and Γ_1 and Γ_2 are the reflection coefficients looking into the left-hand and right-hand short circuits. The values of Γ_1 and Γ_2 depend on the conductivity of the

Table 12.3

Author(s)	Organisation	Year of published paper	Frequency range covered. GHz	Pulse Length ns	Intermediate frequency MHz
King and Mandeville[9]	BTL	1961	33–90	80	70
Berry[10]	Marconi	1965	26·5–40	25	80
Steier[11]	BTL	1965	100–125	–	–
Young and Warters[12]	BTL	1968	50–60	100	70
Shimba[13]	NTT	1969	47–51	–	–
Lacey and Groves[14]	PORD	1970	30–110	200	100
Shimba, Yamaguchi and Kondoh[15]	NTT	1972	47–51	–	–
Moorthy[16]	BTL	1973	85–110	10	300
Bomer[17]	PORD	1974	32–110	1000	100

end walls, the frequency and the tightness of the coupling to the source and receiver that are used for making the Q measurements.

Rearranging eqn. 12.9, we get:

$$\frac{1}{Q} = \frac{\lambda^2}{2\pi\lambda_g}\left(2 - |\Gamma_1| - |\Gamma_2|\right)\frac{1}{l} + \frac{\lambda^2\alpha_1}{\pi\lambda_g} \qquad (12.10)$$

Thus, by plotting $1/Q$ against $1/l$, for several successive resonance points, a straight line is obtained whose intercept on the $1/Q$ axis is

$$\frac{\lambda^2\alpha_1}{\pi\lambda_g}$$

Thus, after finding λ and λ_g, α_1 can be calculated. By adopting this procedure it is not necessary to measure Γ_1 and Γ_2.

Many different techniques have been developed for measuring the Q of a cavity. Sucher and Fox[19] devote 77 pages to this subject and give 41 references. A full treatment of Q measurement techniques is beyond the scope of this book, so the next part of this section will be restricted to brief descriptions of two very interesting methods that have been described recently. The first one was described by Chamberlain[20] and then developed further by Clapham.[21] Fig. 12.7a shows the equipment that is needed to make attenuation measurements when this technique is used to determine the Q values. The sawtooth sweep waveform of the oscilloscope is applied to the reflector of the klystron and the response curve of the cavity, that is formed between the mesh and the sliding piston, is displayed on the oscilloscope. On alternate sweeps, a 1 MHz sinusoidal signal is applied to the reflector as well as the sawtooth waveform. The amplitude of this 1 MHz signal is carefully adjusted until the first-order frequency modulation sidebands have the same amplitude as the residual carrier. From frequency modulation theory[22] it is found that the carrier amplitude is proportional to $J_0(\Delta f/f_m)$ and the amplitudes of the first-order sidebands are proportional to $J_1(\Delta f/f_m)$, where J_0 and J_1 denote Bessel functions of the first kind of zero order and first order, respectively, Δf represents the peak frequency deviation and f_m denotes the modulation frequency. To achieve the condition mentioned earlier, we require that

$$J_0\left(\frac{\Delta f}{f_m}\right) = J_1\left(\frac{\Delta f}{f_m}\right) \qquad (12.11)$$

This equation is satisfied when $\Delta f/f_m = 1\cdot435$ and, with this value, the carrier and the two first-order sidebands have amplitudes that are $5\cdot23$ dB below the amplitude of the unmodulated carrier.

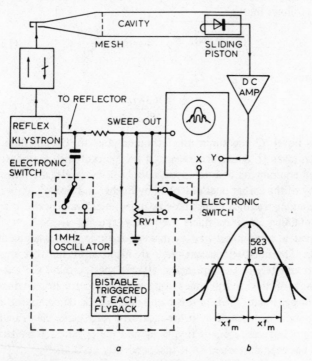

Fig. 12.7 *Q* measurement technique of Chamberlain and Clapham
  a Equipment needed
  b Oscilloscope display when adjustments have been completed

On the alternate sweeps, when the 1 MHz modulation is present, three response curves with 1 MHz separations, appear on the oscilloscope.

During the unmodulated sweeps, the full sawtooth waveform is applied to the X amplifier. During the other sweeps, this sawtooth waveform is reduced by a precision potentiometer RV1 to a value which makes the sideband peaks lie on the unmodulated response curve as shown in Fig. 12.7*b*.

For a single tuned circuit, with a resonant frequency f_0 and a loaded magnification factor denoted by Q, it is well known that

$$\frac{\text{response at resonance}}{\text{response at frequency } \delta f \text{ away from resonance}} \approx \sqrt{1 + 4Q^2 \left(\frac{\delta f}{f_0}\right)^2} \quad (12.12)$$

Let x denote the fraction of the full sawtooth waveform that is applied to the X amplifier, during the sweeps when the modulation is present. Then, it follows that

$$\sqrt{1 + 4Q^2 \left(\frac{xf_m}{f_0}\right)^2} = \text{antilog} \frac{5 \cdot 23}{20} \quad (12.13)$$

From eqn. 12.13, we get

$$Q = \frac{0 \cdot 7638 f_0}{xf_m} \quad (12.14)$$

This novel Q measurement technique due to Chamberlain and Clapham gives Q values accurate to a few percent with a modest amount of equipment, and it appears to be a more elegant technique than any of the earlier ones described by Sucher and Fox.[19]

By making use of a computer-controlled microwave network analyser containing an ultra-stable frequency synthesiser, Skilton[23] has developed a very quick way of making accurate Q measurements at 7·3 GHz. Transmission measurements are made on the network analyser at approximately 100 frequencies, 50 kHz apart, around the resonant frequency of the cavity under test, and the results are obtained on punched paper tape. This tape is then fed into a digital computer which fits a polynomial equation to the measured values and then calculates and prints out the resonant frequency and the Q value. A similar procedure has been developed quite independently by Uhlir.[24]

Beck and Dawson,[25] Paghis[26] and Thorp[27] have used Q measurement techniques to obtain values for the conductivities of various metals at microwave frequencies. Several workers[20, 21, 28–35] have deduced attenuation values for low-loss circular waveguides from Q measurements and Skilton[23] has determined the losses in various types of waveguide connectors by making Q measurements on cavities containing up to 11 flanged couplings.

12.6 Determination of attenuation with a shorted slotted line

Weber[36] has described a very simple method of determining attenuation, but it only gives moderate accuracy. The equipment needed is shown in Fig. 12.8a. The unknown, with an attenuation of A dB, is inserted between isolators that are matched by tuners T1 and T2 and the source seen by the slotted line is matched by tuner T3. A short circuit with the lowest possible loss is connected to the right-hand end of the slotted line. If all losses in the slotted line and short circuit are neglected, then it follows from elementary standing wave theory that the resultant amplitude at distance l from the short circuit is given by

$$E_l = \left| 2 E_1 \sin \frac{2\pi l}{\lambda_g} \right| \qquad (12.15)$$

where E_1 is the incident wave amplitude, and λ_g is the guide wavelength.

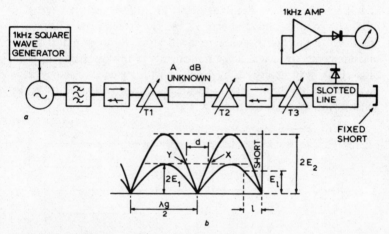

Fig. 12.8 Determination of attenuation with a shorted slotted line
    a Equipment needed
    b Voltage distribution along slotted line

The carriage on the slotted line is moved along until the resultant amplitude at the probe has its maximum value of $2E_1$ (see Fig. 12.8*b*). The unknown is now removed and the two connectors at the insertion point are joined directly together. This causes the incident wave amplitude to rise to E_2 and the resultant amplitude at the probe to increase to $2E_2$, where E_1, E_2 and A are, by definition related as follows:

$$A = 20 \log_{10} \frac{E_2}{E_1} \qquad (12.16)$$

The carriage is now moved successively to the two points X and Y, distance d apart, at which the resultant amplitude at the probe is again $2E_1$. The distance d and the guide wavelength λ_g are measured as accurately as possible. It now follows from eqn. 12.15 and Fig. 12.8*b* that:

$$2E_1 = 2E_2 \sin\frac{2\pi}{\lambda_g} \cdot \frac{d}{2} \qquad (12.17)$$

From eqns. 12.16 and 12.17, we get

$$A = 20 \log_{10}\frac{1}{\sin{(\pi d/\lambda_g)}} \qquad (12.18)$$

This equation is only approximate because all losses in the slotted line and short circuit have been neglected. Fig. 12.9 shows the variation of A with d/λ_g according to eqn. 12.18. The detector law is unimportant because the signal levels are identical at the three points where measurements are taken. Some refinements to this measurement technique have been described, by Hook.[37] With a grade 1 standing wave indicator, he has shown that an accuracy of ± 0.05 dB at 10 dB can be achieved in waveguide sizes 16 and larger.

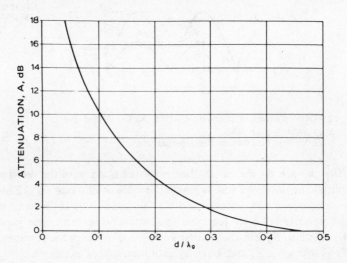

Fig. 12.9 Attenuation values obtained from voltage distribution

The probe penetration must be kept small to minimise probe reflection errors. The best check is to take sets of readings with decreasing probe penetrations and then operate in the region where further reduction of the probe penetration has no significant effect on the results.

12.7 Somlo's fixed-probe method for measuring localised waveguide attenuation

In 1969, Somlo[38] described a simple technique for measuring the attenuation constant of a waveguide as a function of position along the guide. The technique makes it possible to investigate the effects of localised surface roughness, the quality of internal electroplating etc. The equipment needed is shown in Fig. 12.10. The source is carefully matched by tuner T3. The probe is fixed in position and connected to a sensitive receiver. A non-contacting sliding short circuit is moved through the section of waveguide under investigation and three positions are located, the first where the wave amplitude at the probe reaches a maximum value E_{max}, the second $\lambda_g/4$ closer to the source where the wave amplitude at the probe has a minimum value $E_{min,1}$ and the third $\lambda_g/4$ beyond the maximum position where the wave amplitude at the probe has a minimum value $E_{min,2}$.

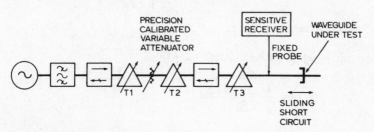

Fig. 12.10 Equipment needed for Somlo's fixed probe method of measuring localised waveguide attenuation

In the first instance, let us assume that the incident wave amplitude has a constant value E_i. Let l denote the distance from the probe to the sliding short circuit when the output is maximum, and let $|\Gamma|$ denote the modulus of the reflection coefficient of this sliding short. Then, from elementary standing wave theory we get

$$E_{max} = E_i \{1 + |\Gamma| e^{-2\alpha_m l}\} \qquad (12.19)$$

$$E_{min,1} = E_i \{1 - |\Gamma| e^{-2\alpha_m l} e^{\alpha \lambda_g/2}\} \qquad (12.20)$$

$$E_{min,2} = E_i \{1 - |\Gamma| e^{-2\alpha_m l} e^{-\alpha \lambda_g/2}\} \qquad (12.21)$$

where α_m is the mean value of the attenuation constant over the distance l and α is the attenuation constant in the immediate vicinity of the sliding short circuit. Both α_m and α are in nepers/cm.

To avoid errors caused by departure from square law in the detector connected to the probe, the precision variable attenuator shown in Fig. 12.10 is used to adjust the incident wave amplitude E_i, so that the resultant wave amplitude at the probe has the same value E_p for each of the three sliding short positions mentioned earlier. Let the three attenuator settings, in dB, that correspond to these three positions be denoted by A_{max}, $A_{min,1}$ and $A_{min,2}$. To simplify the mathematics, let y represent $|\Gamma| e^{-2\alpha_m l}$. Then, denoting the wave amplitude at the variable attenuator input by E_{in}, we get

$$E_p = E_{in} \, 10^{-A_{max}/20} \, (1+y) = E_{in} \, 10^{-A_{min,1}/20} \, (1 - y \, e^{\alpha\lambda_g/2})$$

$$= E_{in} \, 10^{-A_{min,2}/20} \, (1 - y \, e^{-\alpha\lambda_g/2}) \tag{12.22}$$

With slight rearrangement, we now get

$$10^{(A_{max} - A_{min,1})/20} = \frac{1+y}{1 - y \, e^{\alpha\lambda_g/2}} \tag{12.23}$$

and

$$10^{(A_{min,2} - A_{min,1})/20} = \frac{1 - y \, e^{-\alpha\lambda_g/2}}{1 - y \, e^{\alpha\lambda_g/2}} \tag{12.24}$$

Further rearrangement gives

$$\frac{\text{antilog} \dfrac{(A_{min,2} - A_{min,1})}{20} - 1}{\text{antilog} \dfrac{A_{max} - A_{min,1}}{20}} = \frac{y \, (e^{\alpha\lambda_g/2} - e^{-\alpha\lambda_g/2})}{1+y} \tag{12.25}$$

As

$$\alpha\lambda_g/2 \ll 1, \qquad\qquad e^{\alpha\lambda_g/2} \approx 1 + \frac{\alpha\lambda_g}{2}$$

and

$$e^{-\alpha\lambda_g/2} \approx 1 - \frac{\alpha\lambda_g}{2}$$

Also, y is very close to unity; thus, we can replace $y/(1+y)$ by $\frac{1}{2}$. Furthermore, the attenuation constant A, in dB/cm, is $8 \cdot 686 \, \alpha$ (see eqn. 1.3), so we finally get

$$A = \frac{17 \cdot 372}{\lambda_g} \left\{ \frac{\text{antilog} \dfrac{A_{min,2} - A_{min,1}}{20} - 1}{\text{antilog} \dfrac{A_{max} - A_{min,1}}{20}} \right\} \tag{12.26}$$

As an example, if $A_{max} = 40\,\text{dB}$, $A_{min,1} = 0\,\text{dB}$, $A_{min,2} = 1\,\text{dB}$ and $\lambda_g = 4\,\text{cm}$, it is seen from eqn. 12.26 that $A = 0 \cdot 0053\,\text{dB/cm}$.

Somlo[38] has given a nomogram based on eqn. 12.26 which can be used to determine values for the attenuation constant very rapidly.

As is usual when taking v.s.w.r. measurements, the probe penetration must be kept small. The checking procedure described at the end of Section 12.6 is equally valid for this fixed probe technique.

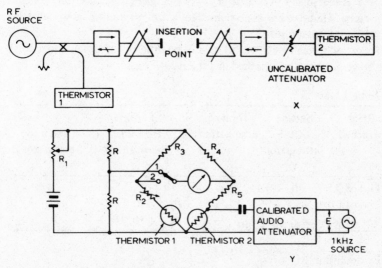

Fig. 12.11 Fulford and Blackwell's attenuation measuring system

12.8 Attenuation measurement with two thermistors in adjacent arms of a wheatstone bridge

Fulford and Blackwell[39] have described an attenuation measurement technique which differs sufficiently from the usual power ratio method to warrant special mention. Fig. 12.11 shows the equipment needed for this technique. The measurement procedure is as follows:

(a) With the connectors at the insertion point joined directly together, the switch in position 1, the r.f. turned on and set to its normal level, the uncalibrated attenuator X set to give minimum attenuation and the 1 kHz oscillator switched off, R_1 is varied until the bridge balances.

(b) The switch is now moved to position 2 and R_2 is varied until the bridge again balances.

(c) Thermistor 2 is disconnected from the microwave circuit, so that it receives no r.f. input and the precision l.f. calibrated attenuator Y is set to zero. The 1 kHz oscillator is now turned on and its output voltage E is adjusted until a balanced condition is obtained for the third time.

(d) Thermistor 2 is reconnected, a known attenuation of $A_{LF,1}$ dB is inserted in Y and balance is then restored by adjustment of X to a value of $A_{RF,1}$. Let the r.f. power that is now fed into thermistor 2 be denoted by P_1.

(e) The unknown, with an attenuation of A_u dB, is now inserted and Y is readjusted to a value $A_{LF,2}$ which gives a balance for the fifth time. Under these conditions, let P_2 denote the r.f. power that is fed into thermistor 2.

To provide greater clarity, all of the essential information about this 5 stage process is summarised in Table 12.4.

Table 12.4

Stage reached	Setting of L.F. attenuator Y	Total r.f. attenuation	L.F. power fed into thermistor 2	R.F. power fed into thermistor 2
End of (c)	0	∞	P_{LF}	0
End of (d)	$A_{LF,1}$	$A_{RF,1}$	$P_{LF} \times 10^{-A_{LF,1}/10}$	P_1
End of (e)	$A_{LF,2}$	$A_{RF,1} + A_u$	$P_{LF} \times 10^{-A_{LF,2}/10}$	P_2

By definition,

$$A_u = 10 \log_{10} \frac{P_1}{P_2} \qquad (12.27)$$

The sum of the l.f. and r.f. powers fed into thermistor 2 must be the same at the end of each stage. Thus, we have

$$P_{LF} = P_{LF} \times 10^{-A_{LF,1}/10} + P_1 = P_{LF} \times 10^{-A_{LF,2}/10} + P_2$$

From eqns. 12.27 and 12.28 we get $\qquad (12.28)$

$$A_u = 10 \log_{10} \frac{1 - 10^{-A_{LF,1}/10}}{1 - 10^{-A_{LF,2}/10}} \qquad (12.29)$$

The use of two thermistors in adjacent arms of the bridge can give increased stability against variations in both the ambient temperature and the r.f. power from the source.

Fulford and Blackwell[39] have estimated that the maximum possible uncertainty for their system is ± 0·007 dB. However, Raff and Sorger[40] have shown, more recently, that a subtle error can be present when a power measuring bridge is supplied simultaneously with d.c. and a.f. power, because there may be an appreciable resistance variation during each audio cycle.

12.9 Method using a Josephson junction in a superconducting loop

Certain metals such as indium, lead, niobium and tin make a transition into a superconductive state when cooled below a few Kelvins. They are then called superconductors. In 1962, Josephson[41] predicted that interesting effects would occur if two superconductors were separated by a very thin insulating layer and direct current passed through them from a high impedance source. Under these conditions he said there would be a zero potential difference across the junction until the current rises to a critical value I_c, in the region of $5\,\mu$A, and then a direct voltage V_B would appear across the junction and a weak oscillation would be generated with a frequency proportional to V_B (483·6 MHz per μV). Josephson's predictions were soon proved experimentally. The Josephson effects have now found many applications.[42] Josephson junctions are now used in ultra-sensitive magnetometers and galvanometers, primary voltage standards, voltage tunable oscillators, infra-red detectors, microwave power and attenuation measuring equipments etc.

A superconducting loop containing a Josephson junction is called a SQUID – Superconducting Quantum Interference Device (see Appendix 8). The current that flows round such a device depends on the magnetic flux Φ linking the loop[43] and, in Appendix 8, it is shown that it is given by

$$I_j = I_c \sin (2\pi\Phi/\Phi_0) \qquad (12.30)$$

I_c depends on the nature of the junction and

$$\Phi_0 = h/2e = 2\cdot0678538 \times 10^{-15} \text{ weber} \qquad (12.31)$$

where h is Planck's constant and e is the charge on an electron. Let Φ_a denote the component of Φ which is due to an external source. Then

$$\Phi = \Phi_a + LI_j \qquad (12.32)$$

where L is the inductance of the loop. From eqns. 12.30 and 12.32, we get

$$I_j = I_c \sin \{2\pi(\Phi_a + LI_j)/\Phi_0\} \qquad (12.33)$$

If we add any integral multiple of Φ_0 to Φ_a, the value of I_j is unchanged. The e.m.f. around the loop is $d\Phi/dt$. Thus, the SQUID behaves like a non-linear impedance which is a periodic function of Φ_a.

Fig. 12.12 shows how Kamper *et al*[44-49] use a SQUID to measure attenuation at 30 MHz. A weak 9 GHz signal is used to indicate the periodic impedance variations of the SQUID as the 30 MHz attenuator under test is changed. A point-contact niobium Josephson junction is formed across a section of very low impedance X-band waveguide that is short circuited $\lambda_g/4$ behind the junction and lined with Babbitt metal (an alloy of tin, lead, antimony and copper). The SQUID thus formed is inductively coupled to the inner conductor of a $50\,\Omega$ coaxial line. The 30 MHz output current from the attenuator under test, a d.c. bias current and either a 1 kHz square wave or a 1 kHz sine wave are passed through this inner conductor.

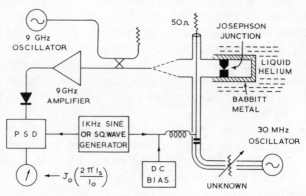

Fig. 12.12 Josephson attenuator calibrator

In the early work on this subject, the 9 GHz power reflected by the SQUID was assumed to be given by the following equation:

$$P = P_0 + P_1 \cos \frac{2\pi I}{I_0} \tag{12.34}$$

where P_0 and P_1 are constants, I is the total current through the inner conductor and I_0 is the current through the inner conductor that is needed to change the flux linking the SQUID by Φ_0.

Let the 30 MHz component of the current through the inner conductor be denoted by $I_s \sin \omega_s t$.

When the 1 kHz signal has a square waveform, its peak-to-peak amplitude is adjusted to a value of $I_0/2$ and the d.c. bias current is set to a value of $I_0/4$.

With these conditions, the resultant amplitude of the combined square wave and bias currents is zero throughout each negative half-cycle of the square wave and $I_0/2$ throughout each positive halfcycle.

Thus, on the negative half cycles, the reflected power is given from eqn. 12.34 by

$$P^- = P_0 + P_1 \cos \frac{2\pi I_s \sin \omega_s t}{I_0} \qquad (12.35)$$

and on the positive half cycles, it is given by

$$P^+ = P_0 + P_1 \cos \frac{2\pi (I_s \sin \omega_s t + I_0/2)}{I_0} \qquad (12.36)$$

$$= P_0 + P_1 \cos \left\{ \frac{2\pi I_s \sin \omega_s t}{I_0} + \pi \right\} \qquad (12.37)$$

$$= P_0 - P_1 \cos \frac{2\pi I_s \sin \omega_s t}{I_0} \qquad (12.38)$$

Let the detector that follows the 9 GHz amplifier have a square law characteristic, and let the overall gain of the 9 GHz amplifier and detector be denoted by G_1. Then the mean output voltage during each negative half cycle of the square wave is

$$V^- = G_1 \left\{ P_0 + \frac{P_1}{2\pi} \int_0^{2\pi} \cos \frac{2\pi I_s \sin \omega_s t}{I_0} \, d(\omega_s t) \right\} \qquad (12.39)$$

$$= G_1 \left\{ P_0 + P_1 \cdot J_0 \left(\frac{2\pi I_s}{I_0} \right) \right\} \qquad (12.40)$$

where J_0 denotes a zero-order Bessel function of the first kind.[50] Similarly, the mean output voltage during each positive half cycle of the square wave is found to be

$$V^+ = G_1 \left\{ P_0 - P_1 J_0 \left(\frac{2\pi I_s}{I_0} \right) \right\} \qquad (12.41)$$

A phase sensitive detector (PSD) gives an output voltage which is proportional to the difference between its mean input voltages on alternate half cycles. Thus, the PSD output is given by

$$V_{\text{out}} = 2 G_1 K_1 P_1 J_0 \left(\frac{2\pi I_s}{I_0} \right) \qquad (12.42)$$

where K_1 depends on the gain in the PSD.

More detailed studies of the Josephson attenuator calibrator revealed that eqn. 12.34 was inadequate, so the 9 GHz power reflected by the SQUID was then represented by a Fourier series and written as

$$P = P_0 + \sum_n P_n \cos \left\{ \frac{2\pi n}{I_0} (I_s \sin \omega_s t + I_m \sin \omega_m t + I_b) \right\} \qquad (12.43)$$

where I_m denotes the peak value of the 1 kHz current, which in this case is assumed to be sinusoidal, and I_b denotes the bias current.

By averaging eqn. 12.43 over a complete r.f. cycle, then using Fourier analysis to find the amplitude of the 1 kHz component in the square law detector output and finally mixing this with the 1 kHz reference signal, the output from the PSD is now found to be[47]

$$V_{out} = G_2 K_2 \sum_n P_n \cdot J_0 \left(\frac{2\pi n I_s}{I_0}\right) \cdot J_1 \left(\frac{2\pi n I_m}{I_0}\right) \sin \left(\frac{2\pi n I_b}{I_0}\right)$$

(12.44)

where G_2 and K_2 are the gains corresponding to G_1 and K_1. These gains will not be identical because of the change in the modulation waveform. In eqn. 12.44, J_1 denotes a first-order Bessel function.

Even harmonic distortion ($n = 2, 4, 6$, etc.) can be completely eliminated by adjusting the bias current until $I_b = 0.25 I_0$ and this condition maximises the desired output signal at $n = 1$. Third harmonic distortion can be eliminated by adjusting the amplitude of the 1 kHz sine wave until $I_m = 0.2033 I_0$. This value makes the first-order Bessel function zero. The fifth- and higher-order odd harmonics can be neglected. Therefore, when I_b and I_m have been correctly set, it follows that

$$V_{out} \propto J_0 \left(\frac{2\pi I_s}{I_0}\right)$$

(12.45)

Thus, both the approximate and more precise treatments show that the PSD output is proportional to $J_0(2\pi I_s/I_0)$. Therefore, it only requires a table of zero-order Bessel function roots to determine the attenuation between any two nulls. The roots of $J_0(x)$ occur at 2.40483, 5.52008, 8.65373 etc.[51]

Let I_{s1}, I_{s2} and I_{s3} denote the first three values of the r.f. current that give a zero output. Then it follows that

$$\frac{2\pi I_{s1}}{I_0} = 2.40483, \frac{2\pi I_{s2}}{I_0} = 5.52008, \frac{2\pi I_{s3}}{I_0} = 8.65373$$

Thus, the change in attenuation needed to go from the first zero to the second is

$$\Delta A_{1-2} = 20 \log_{10} \frac{I_{s2}}{I_{s1}} = 20 \log_{10} \frac{5.52008}{2.40483} = 7.21722 \, dB$$

The change in attenuation needed to go from the second zero to the third is

$$\Delta A_{2-3} = 20 \log_{10} \frac{I_{s3}}{I_{s2}} = 20 \log_{10} \frac{8.65373}{5.52008} = 3.90516 \, dB$$

and so on. To reach 62 dB, one must cover 900 nulls. It is tedious to count them by looking at the output meter, so electronic counters are used to do this.[49] A smooth unidirectional reduction in attenuation is needed to make sure that no nulls are missed.

At NBS, agreement to within ± 0·002 dB over a range of 62 dB was obtained between a SQUID system and the normal calibration service.[49] A similar agreement has been achieved at NPL.[52, 53] Present SQUID systems can be used up to about 1 GHz. With the readout signal at Q-band, it may be possible to carry out attenuation measurements up to X-band.

12.10 Miscellaneous methods

In Chapters 5 to 11 and Sections 12.1 to 12.9, numerous different attenuation measurement techniques have been described, but there are still a few more and these will be described here very briefly.

In 1957, a brief description was given[54] of a self-calibrating 3-channel null system that was first originated in 1953 by C.M. Allred at the National Bureau of Standards. Piston attenuators are used in two of the channels and their outputs are approximately in quadrature. The unknown is inserted in the third channel in series with a phase shifter. The two piston attenuators are adjusted to give nulls at different settings of the phase shifter and a number of simultaneous equations may be obtained. On solving them, both the insertion loss and insertion phase angle of the unknown can be obtained. A further paper[55] related to this technique appeared in 1964.

Lacy and Miller[56] have described an improved method for measuring losses in short waveguide lengths. The waveguide under test is inserted between two identical irises that take the form of $\lambda_g/4$ long reduced height waveguide sections. A cavity is thus formed and the attenuation through it at resonance is measured by an r.f. substitution technique. The attenuation through one of the irises by itself is then measured. From these two results, the loss in the waveguide can be readily calculated. The full theory underlying this technique is given in Lacy and Miller's paper.[56] Losses as small as 0·02 dB can be measured using this technique.

In 1962, Davies[57] described an absolute method of calibrating a variable attenuator using a 3 dB coupler. The attenuator is calibrated in a step-by-step manner in terms of the attenuation A_p through the coupler. The measurements of small attenuation required to correct for the absorption loss and unequal power split of the coupler are made in terms of A_p. Finally, an iterative process is used to determine A_p.

12.11 References

1 LAVERICK, E.: 'The calibration of microwave attenuators by an absolute method', *IRE Trans.*, 1957, **MTT-5**, pp. 250–254
2 PECK, R.L.: 'A method for the self-calibration of attenuation measuring systems', *J. Res. NBS*, 1962, **66C**, pp. 13–18
3 IWASE, T.: 'A new method for small attenuation measurements', *Rep. Electrotech. Lab. (Tokyo)*, 1964, **28**, pp. 13–16
4 NEMOTO, T., BEATTY, R.W. and FENTRESS, G.H.: 'A two-channel off-null technique for measuring small changes of attenuation', *IEEE Trans.*, 1969, **MTT-17**, pp. 396–397
5 LARSON, W., DESCH, R.F. and GILLARD, B.F.: 'Further analysis of the off-null versus power ratio method of attenuation measurement', *IEEE Trans.*, 1970, **MTT-18**, pp. 112–113
6 ENGEN, G.F. and BEATTY, R.W.: 'Microwave attenuation measurements with accuracies from 0.0001 to 0.06 decibel over a range of 0.01 to 50 decibels', *J. Res. NBS*, 1960, **64C**, pp. 139–145
7 NEMOTO, T.: 'The measurements of a waveguide joint properties', *Electron. & Commun. Jap.*, 1969, **52B**, pp. 68–76
8 RING, D.H.: unpublished paper
9 KING, A.P. and MANDEVILLE, G.D.: 'The observed 33 to 90 kmc attenuation of two-inch improved waveguide', *BSTJ*, 1961, **40**, pp. 1323–1330
10 BERRY, J.A.: 'Measurement of H_{01} mode loss at Q-band in helix waveguide', *Marconi Rev.*, 1965, **28**, pp. 22–26
11 STEIER, W.H.: 'The attenuation of the Holmdel helix waveguide in the 100–125 Kmc band', *BSTJ*, 1965, **44**, pp. 899–906
12 YOUNG, D.T. and WARTERS, W.D.: 'Precise 50 to 60 GHz measurements on a two-mile loop of helix waveguide', *BSTJ*, 1968, **47**, pp. 933–955
13 SHIMBA, M.: 'Attenuation measurement of millimeter waveguide by the pulse-reflection method', *Electron. & Commun. Jap.* 1969, **52A**, pp. 53–58
14 LACEY, N. and GROVES, I.S.: 'Measurement of TE_{01} mode attenuation by the shuttle pulse method' *in* 'Trunk telecommunication by guided waves', *IEE Conf. Publ.* 1970, **71**, pp. 137–141
15 SHIMBA, M., *et al.*: 'Method of measuring millimeter waveguide line attenuation', *IEEE Trans.*, 1972, **IM-21**, pp. 215–219
16 MOORTHY, S.C.: 'A computerized system for the measurement of attenuation in a helix waveguide', *IEEE Trans.*, 1973, **IM-22**, pp. 311–314
17 BOMER, R.P.: 'A computer controlled attenuation measurement system for TE_{01} mode circular waveguide from 32–110 GHz', *IEEE Trans.*, 1974, **IM-23**, pp. 386–389
18 BARLOW, H.M. and CULLEN, A.L.: 'Microwave measurements' (Constable, 1950), chap. 3
19 SUCHER, M. and FOX, J.: 'Handbook of microwave measurements – Vol. 2' (Polytechnic Press,, 1963), chap. 8

20 CHAMBERLAIN, J.K.: 'Q measurements on low-loss waveguide cavities', *Electron. Engineering*, 1966, **38**, pp. 579–581

21 CLAPHAM, W.J.: 'Measurement of low-loss waveguide attenuation at 85 GHz by a resonance method' *in* 'Trunk telecommunications by guided waves', *IEE Conf. Publ.* 1970, **71**, pp. 303–306

22 GOLDMAN, S.: 'Frequency analysis, modulation & noise' (McGraw-Hill, 1948), chap. 5

23 SKILTON, P.J.: 'A technique for determination of loss, reflection, and repeatability in waveguide flanged couplings', *IEEE Trans.*, 1974, **IM-23**, pp. 390–394

24 UHLIR, A.: 'Automatic microwave Q measurement for determination of small attenuations', *IEEE Trans.*, 1972, **MTT-20**, pp. 38–41

25 BECK, A.C., and DAWSON, R.W.: 'Conductivity measurements at microwave frequencies', *Proc. IRE*, 1950, **38**, pp. 1181–1189

26 PAGHIS, I.: 'Surface losses in electromagnetic cavity resonators', *Can. J. Phys.* 1952, **30**, pp. 174–184

27 THORP, J.S.: 'R.F. conductivity in copper at 8 mm wavelengths', *Proc. IEE*, 1954, **101**, Pt. III, pp. 357–359

28 YOUNG, J.A.: 'Resonant-cavity measurements of circular electric waveguide characteristics', *Proc. IEE*, 1959, **106B**, Suppl. 13, pp. 62–65

29 KARBOWIAK, A.E. and SKEDD, R.F.: 'Testing of circular waveguides using a resonant cavity method', *Proc. IEE*, 1959, **106B**, Suppl. 13, pp. 66–70

30 KEITH-WALKER, D.G.: 'An equipment for measuring the attenuation of low-loss waveguide transmission lines', *Proc. IEE*, 1959, **106B**, Suppl. 13, pp. 71–74

31 HAMER, R. and WESTCOTT, R.J.: 'Measurement of TE_{01} mode attenuation in short lengths of circular waveguide', *Proc. IEE*, 1962, **109B**, Suppl. 23 pp. 814–819

32 SOMLO, P.I.: 'Cable attenuation measurement', *Proc. IRE (Australia)*, 1962, **23**, p. 585

33 CHILDS, G.H.L.: 'Measurement of TE_{01} mode attenuation in low-loss waveguide by a decrement method' in 'Trunk telecommunication by guided waves', *IEE Conf. Publ.* 1970, **71**, pp. 269–274

34 OLVER, A.D., CLARRICOATS, P.J.B. and CHONG, S.L.: 'Experimental determination of attenuation in corrugated circular waveguides', *Electron. Lett.*, 1973, **9**, pp. 424–426

35 CLARRICOATS, P.J.B., OLVER, A.D. and CHONG, S.L.: 'Attenuation in corrugated circular waveguides. Part 2: experiment', *Proc. IEE*, 1975, **122**, pp. 1180–1186

36 WEBER, E.: 'The measurement of attenuation', in MONTGOMERY, C.G. (Ed.): 'Technique of microwave measurements' (McGraw-Hill, 1947)

37 HOOK, A.P.: 'Calibrating waveguide attenuators', *Bri. Commun. & Electron.*, 1960, **7**, pp. 922–923

38 SOMLO, P.I.: 'Localized waveguide attenuation measurement', *Proc. IREE (Australia)*, 1969, **30**, pp. 13–15

39 FULFORD, J.A. and BLACKWELL, J.H.: 'Accurate method for measurement of microwave attenuation', *Rev. Sci. Instrum.* 1956, **27**, pp. 956–958

40 RAFF, S.J. and SORGER, G.U.: 'A subtle error in RF power measurements', *IRE Trans.*, 1960, **I-9**, pp. 284–291

41 JOSEPHSON, B.D.: 'Possible new effects in superconductive tunnelling', *Phys. Lett. (Netherlands)*, 1962, **1**, pp. 251–253

42 PETLEY, B.W.: 'An introduction to the Josephson effects' (Mills & Boon, 1971)

43 SILVER, A.H. and ZIMMERMAN, J.E.: 'Quantum states and transitions in weakly connected superconducting rings', *Phys. Rev.*, 1967, **157**, pp. 317–341

44 KAMPER, R.A. and SIMMONDS, M.B.: 'Broadband superconducting quantum magnetometer', *Appl. Phys. Lett.*, 1972, **20**, pp. 270–272

45 KAMPER, R.A. *et al.*: 'Quantum mechanical measurements of RF attenuation', Proceedings of the Applied Superconductivity Conference, pp. 696–700, 1972

46 KAMPER, R.A. *et al.*: 'A new technique for RF measurements using superconductors', *Proc. IEEE*, 1973, **61**, pp. 121–122

47 KAMPER, R.A. *et al.*: 'Measurement of R.F. power and attenuation using superconducting quantum interference devices'. NBS Technical Note 643, August 1973

48 KAMPER, R.A. *et al.*: 'Advances in the measurement of RF power and attenuation using SQUIDS., NBS Technical Note 661, Sept. 1974

49 ADAIR, R.T. *et al.*: 'RF attenuation measurements using quantum interference in superconductors', *IEEE Trans.*, 1974, **IM-23**, pp. 375–381

50 McLACHLAN, N.W.: 'Bessel functions for engineers' (Oxford University Press, 1934), chap. 3

51 British Association Mathematical Tables – Vol. 6 Bessel Functions, Part 1' (Cambridge University Press, 1937) pp. 171–173

52 STEELE, J. McA., DITCHFIELD, C.R., and BAILEY, A.E.: 'Electrical standards of measurement. Part 2: R.F. and microwave standards', *Proc. IEE*, 1975, **122**, pp. 1037–1053

53 PETLEY, B.W., MORRIS, K., YELL, R.W., and CLARKE, R.N.: 'Moulded microwave SQUID for r.f. attenuator calibration', *Electron. Lett.*, 1976, **12**, pp. 237–238

54 'Self-calibrating method of measuring insertion ratio', NBS Tech. News Bull., *1957*, **41**, pp. 132–133

55 ALLRED, C.M. and LAWTON, R.A.: 'Precision detector for complex insertion ratio measuring systems', *IEEE Trans.*, 1964, **IM-13**, pp. 76–81

56 LACEY, P.D. and MILLER, K.E.: 'An improved method for measuring losses in short waveguide lengths', Hewlett Packard J., 1957, **9**, pp. 1–4 and 6

57 DAVIES, M.C.: 'An absolute method for calibrating microwave attenuators', *Proc. IEE*, 1962, **109B**, Suppl. 23, pp. 796–800

Chapter 13

Attenuation transfer standards

13.1 Introduction

To find out how the results obtained with different attenuation measuring systems compare with each other, attenuation transfer standards are needed. An attenuation transfer standard should be very rugged and readily portable, its microwave characteristics should remain extremely stable over a period of many years and it should have a very low v.s.w.r. under all conditions to minimise the mismatch error. It should provide an attenuation change without it being necessary to separate and recouple any microwave connectors. If this is not the case, it should be fitted with extremely high class waveguide flanges or coaxial connectors. It should have a low temperature coefficient and negligible leakage; its characteristics should be totally unaffected by microwave power levels up to at least 100 mW and its input and output ports should ideally be in-line. If a variable attenuator is used as a transfer standard, it should have an expanded scale and negligible backlash. Furthermore, no errors must arise in the mechanism linking the attenuating vane to the scale.

Many different devices can be considered for use as attenuation transfer standards and all of the popular ones will now be discussed.

13.2 Directional coupler transfer standards

A strongly-made directional coupler, in either waveguide or coaxial line, forms a very rugged, readily portable, well-matched, low temperature coefficient transfer standard. However, to measure such a device, the two connectors at the insertion point must first of all be connected directly together (to set the zero of the measuring equipment) and they then have to be moved apart so that the transfer standard can be inserted. To make this possible, it is necessary to move either the source or the receiver, or have a length of flexible waveguide or flexible coaxial cable in the measuring equipment. Horizontal movement along the axis of the transmission system is the easiest manoeuvre that can be carried out, so it is desirable to have an inline transfer standard. Unfortunately conventional waveguide and coaxial directional couplers do not satisfy this requirement. However, there are many different ways in which directional couplers can be made so that their input and output ports become inline with each other.[1] Five possibilities are shown in Fig. 13.1. Another solution, for attenuation values below 3 dB, is to use the main arm of a directional coupler as a transfer standard (see Fig. 13.2). Larson[1] has found that the spread in a set of attenuation measurements on a true inline directional coupler is distinctly lower than that for a conventional directional coupler.

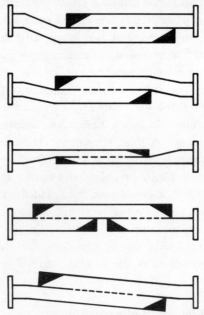

Fig. 13.1 Five different in-line directional couplers

13.3 Directional coupler waveguide switch transfer standards

Extremely good waveguide attenuation transfer standards can be made by combining directional couplers with repeatable waveguide switches. The simplest arrangement of this type is shown in Fig. 13.3a. The earliest use of this configuration appears to have been in an Australian radiometer.[2] In one position of the switch, the input port is connected straight through to the output port. In the other switch position, the input port is connected to the main arm of the directional coupler and the power that is coupled into the side arm is fed through the other arc of the switch to the output. A unit of this type places great demands on the reproducibility of the waveguide switch. Hundreds of measurements made at RRE on Decca and Midcentury waveguide switches have shown that they are repeatable to better than 0·001 dB. A more versatile coupler switch transfer standard is shown in Fig. 13.3b With the coupling coefficients shown on the diagram, this arrangement provides attenuation values close to 0, 10, 20 and 30 dB. Many other variations on this theme are possible. As a final example, Fig. 13.4 shows how four waveguide switches can be combined with four directional couplers to form a transfer standard which covers the range 0 to 75 dB in 5 dB steps.

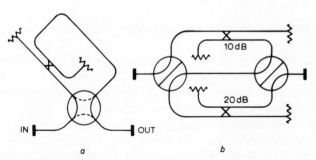

Fig. 13.2 Diagram showing how the main arm of a directional coupler can be used as an attenuation transfer standard

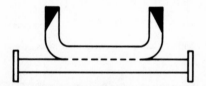

a   b

Fig. 13.3 Directional coupler/waveguide switch transfer standards

a Simplest
b More versatile arrangement

A directional coupler waveguide switch transfer standard is not affected by connector losses and it can be measured a large number of times very quickly without unbolting or reconnecting any waveguide flanges and without moving about any parts of the attenuation measuring system. The directional coupler can be replaced by a fixed stable waveguide attenuator.

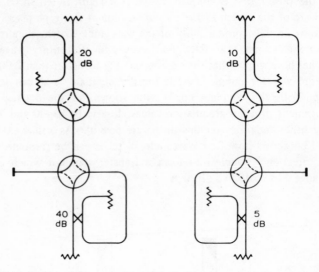

Fig. 13.4 Switched-coupler transfer standard that gives attenuation changes over the range 0 to 75 dB in 5 dB steps

Coaxial equivalents of the configurations shown in Figs. 13.3 and 13.4 are not particularly successful, as coaxial switches are not as repeatable as waveguide switches.

13.4 Side-arm-switched directional coupler transfer standards

A novel type of attenuation transfer standard was described recently by Somlo and Morgan.[3] In this device, the attenuation through the main arm of a directional coupler is changed by switching one of the side ports from a match to a short circuit (see Fig. 13.5). A fixed short circuit is placed on the other side port, at an antiresonant position. With an ideal coupler and both the source and load perfectly matched, the signal flow graph is shown in Fig. 13.6. Using Mason's non-touching loop rule, it is seen at once that the transmission coefficient from port 1 to port 2 is

$$s_{21} = \frac{T(1 - T^2 s e^{j\phi}) + C^2 T s e^{j\phi}}{1 - T^2 s e^{j\phi}} \qquad (13.1)$$

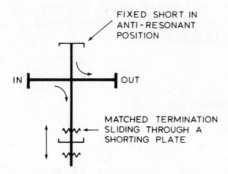

Fig. 13.5 Somlo and Morgan's side-arm-switched directional coupler transfer standard

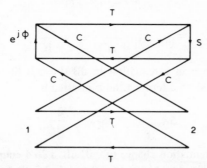

Fig. 13.6 Signal flow graph for the transfer standard shown in Fig. 13.5

For an ideal directional coupler, we have[4]:

$$|T|^2 + |C|^2 = 1 \qquad (13.2)$$

$$T = |T| \, e^{j\psi} \qquad (13.3)$$

$$C = j \, |C| \, e^{j\psi} \qquad (13.4)$$

From eqns. 13.1 to 13.4, we get

$$s_{21} = \frac{|T| \, e^{j\psi} \, \{1 - s \, e^{j(2\psi + \phi)}\}}{1 - |T|^2 s \, e^{j(2\psi + \phi)}} \qquad (13.5)$$

To create a short-to-match switch, Somlo and Morgan[3] placed a metallic partition across the middle of the waveguide, parallel to the electric vector. Since the two halves of the waveguide are below cut-off, full reflection results, so, in this case, we can write

$$s = e^{j\xi} \tag{13.6}$$

To achieve the conversion to a match, they pass a sliding termination with a central slot beyond this partition, thereby making $s = 0$. Substituting eqn. 13.6 in 13.5 we get

$$s_{21}{}^{(1)} = \frac{|T| e^{j\psi} \{1 - e^{j(2\psi + \phi + \xi)}\}}{1 - |T|^2 e^{j(2\psi + \phi + \xi)}} \tag{13.7}$$

The path lengths are chosen so that

$$e^{j(2\psi + \phi + \xi)} = -1 \tag{13.8}$$

giving

$$s_{21}{}^{(1)} = \frac{2|T| e^{j\psi}}{1 + |T|^2} \tag{13.9}$$

In the matched case when $s = 0$, it follows from eqn. 13.5 that

$$s_{21}{}^{(0)} = |T| e^{j\psi} \tag{13.10}$$

Thus

$$\frac{s_{21}{}^{(1)}}{s_{21}{}^{(0)}} = \frac{2}{1 + |T|^2} = \frac{2}{2 - |C|^2} \tag{13.11}$$

Hence, the change in attenuation in dB through the main arm that occurs when the slotted termination is inserted is

$$\Delta A_{sm} = 20 \log_{10} \frac{2}{2 - |C|^2} \tag{13.12}$$

To obtain an attenuation change of 0·01 dB, a 26·4 coupler is required. With a 3 dB coupler the attenuation change is 2·5 dB. In the latter case, there is quite a large change in the input reflection coefficient (from 0 to 0·33), but, for coupling coefficients greater than 20 dB, the change in the input reflection coefficient is very small. Thus, this type of transfer standard is admirably suitable for producing small, very stable attenuation changes. Furthermore, it is seen from eqn. 13.11 that no phase change occurs, at the design frequency, when the slotted termination is inserted.

13.5 Variable waveguide attenuator transfer standards

A well-designed sideways displacement attenuator is suitable for use as a transfer standard. An attenuator of this type is shown in Fig. 13.7. The attenuation is varied by moving the lossy vane sideways across the waveguide and its value is given approximately by[5]

$$A = A_0 + A_v \sin^2\left(\frac{\pi x}{a}\right) \qquad (13.13)$$

where A_0 is the residual attenuation and $A_0 + A_v$ is the total attenuation in dB when the vane is in the centre of the waveguide. Eqn. 13.13 is only approximate because the attenuator law is affected by stray capacitance coupling between the vane and the waveguide walls. On differentiating eqn. 13.13, dA/dx is found to be zero when the vane is at either side or in the centre. Thus, precise positioning of the vane is not essential if it is just used to provide one incremental value of attenuation A_v by moving the vane from one side to the centre. With a metallised glass vane, the temperature coefficient is very low[6], but the value of A_v varies with frequency.

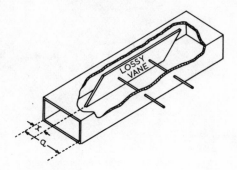

Fig. 13.7 Sideways displacement attenuator

A precision rotary vane attenuator forms a good variable attenuation transfer standard. Commercially-available rotary vane attenuators are not really adequate for this task due to various shortcomings which have already been discussed in Chapter 4. A precision rotary vane attenuator containing a very accurate optical angular measuring system[7,8] would make an excellent transfer standard, but such a device is expensive. Foote and Hunter[9] have improved the resetting accuracy of a commercially available rotary attenuator by a factor of 20 by fitting it with a 180:1 precision Spiroid gear set.

At RRE, a series of special rotary vane attenuators has been developed for use as transfer standards.[10] For reasons given in Chapter 4, key features of the design were: "No worm gears and no end stops". The electroformed waveguides and glass vanes for them were obtained from Flann Microwave Instruments Ltd. of Bodmin, and they were fitted into RRE purpose-designed precision antibacklash spur gear boxes that were manufactured and built in the RRE Engineering Unit. An X-band version is shown in Fig. 13.8.

For each waveguide size, the train of ultra-precise spur gears gives a reduction of 3600:1 in four steps, and it can be driven in either direction by a small permanent magnet d.c. motor or turned manually.

Fig. 13.8 X-band rotary attenuation transfer standard developed at RRE

Each pinion in the gear train is cut integral with its stainless steel shaft and meshes with an aluminium alloy antibacklash gear which is designed with a split hub and clamped to its respective shaft. The total backlash in the gear box is controlled, by careful choice of the anti-backlash spring strengths, to be not more than $0.001°$ at any point throughout the $90°$ movement of the rotor. All shafts are mounted in precision flanged instrument ball bearings with ABEC7 tolerances and are end-loaded to prevent shake and radial movement. The rotor is mounted in extra-low radial runout angular contact ball bearings that again have ABEC7 tolerances and are also end-loaded. The angular position of the rotor is read from three drum dials that are attached to appropriate shafts in the drive train and are graduated, respectively, in $10°$, $0.1°$ and $0.001°$ steps. The effective scale length is 180 m whereas the longest scale length fitted to a British commercially-available rotary attenuator is only 2·4 m. There is no restriction to the angular movement so it is possible to select the most accurate quadrant for normal operation.

Up to now, 13 precision rotary attenuators of this type have been completed in British waveguide sizes 10, 11A, 14, 15, 16, 18 and 22. Backlash and resettability measurements have been made on all of them using an optical polygon and an electronic autocollimator. Fig. 13.9 shows a typical curve of the backlash plotted against the angle turned through to reach a specified dial reading. In this particular case, it is seen that the backlash does not exceed $0 \cdot 0002°$. Hundreds of microwave measurements made on them, using the RRE modulated subcarrier systems, have agreed very well with the autocollimator results and have confirmed that both the backlash and resetting accuracy are less than $0 \cdot 001°$ under all conditions.

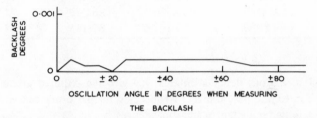

Fig. 13.9 Backlash measurements done on *J*-band precision rotary attenuator 2 on 5/1/1972

Although absolute accuracy is not essential in a transfer standard, some of these precision rotary attenuators have followed the theoretical $40 \log_{10}(\sec \theta)$ law to within $\pm 0 \cdot 006$ dB up to 40 dB.

13.6 Coaxial attenuator transfer standards

Fixed coaxial attenuators are frequently used as attenuation transfer standards. An attenuator of this type sometimes consists of T or π sections (see Fig. 13.10) made with rod and disc resistors, which are mounted in a coaxial line.[11] When a single T or π section is required to give an attenuation of A dB, at a low frequency, and be matched to a line of characteristic impedance Z_0, the design equations are readily found to be:

$$R_1 = Z_0 \frac{1-k}{1+k} \qquad R_2 = 2Z_0 \frac{k}{1-k^2} \qquad (13.14),$$
$$(13.15)$$

$$R_3 = Z_0 \frac{1+k}{1-k} \qquad R_4 = Z_0 \frac{1-k^2}{2k}$$
$$(13.16), (13.17)$$

$$k = \frac{1}{\text{antilog}\,(A/20)} \qquad (13.18)$$

Table 13.1   typical resistance values when $Z_0 = 50\ \Omega$

	T section		*π* section	
A	R_1	R_2	R_3	R_4
dB	Ω	Ω	Ω	Ω
1	2·8751	433·34	869·55	5·7692
2	5·7312	215·24	436·21	11·615
3	8·5499	141·93	292·40	17·615
6	16·614	66·931	150·48	37·352
10	25·975	35·136	96·248	71·151
20	40·909	10·101	61·111	247·50

T and *π* sections designed in this way give the required attenuation from d.c. to h.f., but, at higher frequencies, the performance is affected by distributed inductances and capacitances.

In coaxial attenuators intended for use at frequencies above 1 GHz, either a thin resistive film is deposited on the inner conductor, or a resistive film is deposited on a stripline substrate which is furnished with suitable launching, receiving and earth electrodes.[12]

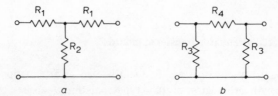

Fig. 13.10 Attenuator sections
   a One section of a *T*-type attenuator
   b One section of a *π*-type attenuator

High quality connectors (eg. GR900 or APC-7) are needed on a coaxial attenuation transfer standard, and it is important to make sure that they are in good mechanical condition, clean and correctly aligned before each mating. Also, they should always be tightened up with a suitable torque spanner. After taking all of these precautions, connector repeatability may still be the largest source of uncertainty in a coaxial attenuation intercomparison (see Table 5.1).

Coaxial attenuators are small rugged inline devices, and well-designed ones have a temperature coefficient of less than 0.0001 dB/dB/$^{\circ}$C, a power sensitivity of less than 0.0003 dB/dB/W and a v.s.w.r. of less than 1.5 from d.c. to 18 GHz.

13.7 References

1 LARSON, W.: 'Inline waveguide attenuator', *IEEE Trans.*, 1964, **MTT-12**, pp. 367–368

2 SOMLO, P.I., and HOLLWAY, D.L.: 'The Australian national standards laboratory X-band radiometer for the calibration of noise sources', *IEEE Trans.*, 1968, **MTT-16**, pp. 664–669

3 SOMLO, P.I., and MORGAN, I.G.: 'A side-arm-switched directional coupler as a single step attenuator of high long-term stability', *IEEE Trans.*, 1974, **MTT-22**, pp. 830–835

4 KAHN, W.K., and KYHL, R.L.: 'The directional coupler core of an arbitrary lossless reciprocal 4-port', *Proc. IRE*, 1961, **49**, pp. 1699–1700

5 MONTGOMERY, C.G.: 'Technique of microwave measurements', Vol. 11 in the Radiation Laboratory Series (McGraw-Hill, 1947), pp. 790–799

6 WALLIKER, D.A.J.: 'Resistive attenuators and their calibration', *Proc. IEE*, 1962, **109B**, Suppl. 23, pp. 791–795

7 LITTLE, W.E., LARSON, W., and KINDER, B.J.: 'Rotary vane attenuator with an optical readout', *J. Res. NBS*, 1971, **75C**, pp. 1–5

8 WARNER, F.L., WATTON, D.O., and HERMAN, P.: 'A very accurate X-band rotary attenuator with an absolute digital angular measuring system', *IEEE Trans.*, 1972, **IM-21**, 446–450

9 FOOTE, W.J., and HUNTER, R.D.: 'Improved gearing for rotary-vane attenuators', *Rev. Sci. Instrum.*, 1972, **43**, pp. 1042–1043

10 WARNER, F.L., WATTON, D.O., HERMAN, P., and CUMMINGS, P.: 'Automatic calibration of rotary vane attenuators on a modulated sub-carrier system', *IEEE Trans.*, 1976, **IM-25**, pp. 409–413

11 ELLIOTT, J.S.: 'Coaxial attenuation standards', *Bell Lab. Rec.*, 1949, **27**, pp. 221–225

12 LUSKOW, A.A.: 'Precision attenuators for microwave frequencies', *Marconi Instrum.*, 1974, **14**, pp. 81–86

Chapter 14

Elimination of error and determination of uncertainty

14.1 Introduction

In microwave attenuation measurements and standards the elimination of error and determination of uncertainty are seldom simple tasks and their importance requires that they should form one of the main objectives of the work.

To make the subject meaningful, it is imperative to start with a set of precise definitions of the properties to be measured, which should be part of a coherent system with the various quantities interconnected by well-established simple mathematical relations. This can be based on the work in Chapter 2 from the definitions and connections between reflection coefficients and attenuation, according to microwave network theory.

Out of this arises the need to define the network properties of single-port or multi-port microwave components relative to precisely defined cross-sections, if these properties are to depend only on the component in question. To meet this need the so-called 'precision' coaxial connectors and the corresponding 'chokeless' connectors (or flanges) for waveguides have been developed. With these, rigid coaxial lines or waveguides can be connected and disconnected at well-defined cross-sections and have a substantially uniform characteristic impedance near the connection, with closely repeatable results. Ordinary coaxial connectors have male and female versions, and the outer and inner conductors usually connect by sprung sleeve action leaving only a vaguely-located cross-section of connection. Ordinary waveguide connectors have 'choke' and 'plain' versions, again with a less well-defined plane of connection than with a precision waveguide connector-pair. The effect of using ordinary connectors at the critical cross-sections (reference planes) in measurement work is an imprecise realisation of properties to be measured, resulting in a systematic uncertainty additional to that otherwise present in the results. In any case, whether precision or ordinary connectors are used, *small* reflection coefficients are often specified as a maximum magnitude over a given frequency range, with no information on phase angles. The effect of any multiple reflections then appears as a small systematic uncertainty with a well-defined range. Such types of uncertainty will be considered later.

14.2 Errors and uncertainties

Error is defined as the measured value minus the correct value. The correct value can usually be obtained (to within a tolerance) by calibration against standards, and in the case of national standards can be inferred from international comparisons, but always with a given uncertainty. This inevitable uncertainty gives rise to much of what follows.

An error that remains unchanged when a measurement is repeated under substantially the same conditions, but can be detected when the apparatus or method is suitably altered, is termed a *systematic error*. It is clearly important to try to locate every possible source of such error and then correct it. Thus, once a source of systematic error has been identified, variation of the conditions can, by a general reversal procedure, lead to the removal of the error. Known sources of systematic error associated with the various methods of attenuation measurement are stated in the chapters dealing with them. Errors which are recognised but cannot be removed or evaluated can only be described by an uncertainty (usually specified by a range) and together they form a *systematic component of uncertainty*.

When a simple measurement is repeated under substantially the same conditions, the results will not all be the same unless the instruments used are too insensitive to detect the small but inevitable differences. The spread in the results obtained is a measure of the presence of a *random component of uncertainty*, the amount depending on the apparatus, the method and sometimes on the person who makes the measurement. Random uncertainties are treated and evaluated by the statistical approach to the theory of probability. The most useful assessment of the true value (apart from systematic error) is the arithmetic mean of all the results, if these involve measurement by differences rather than by ratios, or if the variations are small compared with the mean value. Then, the most useful statistic to express a measure of the random spread in results is the *root mean square* uncertainty.

The word error has often been used to describe what we term here random component of uncertainty, and is still so used. This use of the term is respected, but its existence makes it imperative to be very clear which meaning of error is adopted. Thus the word 'error' has been used by different writers with either one or other of two distinct meanings:

(a) the measured value minus the notionally correct value, for which a correction can obviously be applied whenever it is known

(b) a quantity expressing the amount of spread in the results of multiple repetitions of a measurement, or the uncertainty in one's knowledge of any systematic departure from the correct value (determined by different and better methods).

Failure to be consistent in the use of these two different meanings of the word 'error' has led to much confusion and wrong conclusions, particularly in the subject of systematic components of uncertainty. In this book we use the word 'error' to mean (a) and the word 'uncertainty' to mean (b).

14.3 Application of probability

Having corrected the results for identifiable systematic errors in measurements for the establishment of standards, or having corrected the results in accordance with calibration against standards, one is left with systematic and random components of uncertainty. Both components can in general be expressed through *probability density distributions*. The distribution of the random component is derived and interpreted as *statistical* probability, obtained by sampling the relative frequency of occurrence, but the distribution of the systematic component is derived and interpreted in a manner similar to that for *a priori* probability but with as much experimental information as possible. The mathematics of the theory of probability is the same for both statistical and *a priori* foundations.

The probability density distribution of the results of repeated measurements nearly always approximates to the Gaussian or 'normal' form, provided either that the measurement is by differences rather than by ratios, or that the fluctuations about the mean $\bar{x}$ are small so that, if the measurement is a function $f(x)$ of x, then $f(x + \Delta x) = f(x) + f'(x) . \Delta x$ because $(\Delta x)^2$ and higher powers are negligible in the Taylor series. If neither of these conditions is met, the distribution may be skew, and then the arithmetic mean will not be the best estimate of the quantity measured. If either is met, then the normal distribution will result, and this can be regarded as the limiting form of the distribution of a sum of many small independent random variables, each with any reasonable form of distribution with a range that does not vary excessively between them. In practice, a surprisingly small number of independent component uncertainties (e.g. 3 or more) will result in a close approximation to a normal distribution. But it will be truncated between limits separated by the sum of the individual limits. For random components of uncertainty, statistical sampling theory is applied when only a few repetitions of a measurement can be made and realistic estimates of uncertainty are required. Thus the mean is calculated from

$$\bar{x} = \frac{1}{n} \sum_i x_i \qquad (14.1)$$

and the unbiased estimate of variance is

$$s^2 = \frac{1}{n-1} \sum_i (x_i - \bar{x})^2 \qquad (14.2)$$

where n is the number of measurements and x_i are the individual results ($i = 1$ to n). The bounds of the mean result are then

$$\bar{x} \pm ts/\sqrt{n} \qquad (14.3)$$

where t is the 'Student-t' for a given probability that the correct value is within the bounds stated, apart from systematic uncertainty and uncorrected error. This probability is termed the *confidence level*, usually 0·95 to 0·99. t is tabulated as a function of n and the confidence level in Table 14.1 at the end of Section 14.5.

[In the past, a confidence level of 0·5 was used, giving what was termed the 'probable error'; or a confidence level of 0·683 was used, giving the 'standard error'. But these confidence levels appeared to generate over-optimism because the large probability (0·5 and 0·317) of the correct value being outside the bounds was not always appreciated.]

Application of statistical theory is valid only when the measurement is a stationary statistical process, i.e. the repeated results are consistent. A simple test for this is to divide a series of 10 consecutive results (of one measurement) into 2 groups of 5 each. The mean values $\bar{x}_1, \bar{x}_2$ and the variances s_1^2, s_2^2 of the 2 groups are calculated from eqns. 14.1 and 14.2 with $n = 5$. Form the ratio $F_0 = s_1^2/s_2^2$ (where s_1 is chosen from the pair to ensure that $s_1 > s_2$). Then, if $F_0 < 6\cdot4$, the variance is likely to be 'stationary'. If this is so, then calculate

$$t_0 = 2 \cdot 24 \, |\bar{x}_1 - \bar{x}_2| / \sqrt{s_1^2 + s_2^2}$$

and if $t_0 < 2 \cdot 3$, then the mean value is 'stationary'. If $F_0 > 6 \cdot 4$ and/or $t_0 > 2 \cdot 3$, then some condition probably varies in an erratic manner during the repetitions and the measurement is not a good one.

Systematic components of uncertainty can be divided into two groups, the first arising from imperfection in the measurement process or apparatus, and the second arising from uncertainty in the results of previous measurements on which the measurement in question depends. Most of the first group can be removed by calibration against a standard for ordinary measurements and is only a problem for those who establish standards of measurement. Usually, only the bounds $\pm a$ of a systematic uncertainty are estimated, and one asks what form of probability density distribution is to be associated with these bounds. The assumption of a rectangular (uniform) distribution would assume equal *a priori* likelihood of any value between $\pm a$. The most pessimistic assumption would be of discrete values at $\pm a$, each with a probability $\frac{1}{2}$.

Part of the first group arises from certain difficulties in the measurement of phase angle of small reflection coefficients, and the habit of recording only the maximum over a wide frequency range of the moduli of reflection coefficients, for convenience and economy. This will be dealt with in Section 4.

The second group of systematic uncertainties is treated in a way that depends on how the resultant uncertainties in the results of previous measurements have been quoted. The problem is straightforward if they have been quoted as overall uncertainities having a stated confidence level, or as appropriate statements of probability density distributions.

If the results of the earlier measurements $x_1, x_2 \ldots$ are related to the final quantity z by the relation $z = f(x_1, x_2 \ldots)$ and if the variations $\Delta x_1, \Delta x_2 \ldots$ are independent, then that part of the mean square deviation in the final measurement that arises from uncertainty in the earlier measurements is

$$s_z^2 = (\partial f / \partial x_1)^2 \, s_1^2 + (\partial f / \partial x_2)^2 \, s_2^2 + \ldots \tag{14.4}$$

where s_1, s_2 are the r.m.s. deviations associated with the earlier individual measurements.

14.4 Uncertainty arising from systematic error associated with multiple reflections

In microwave attenuation measurements there is a systematic error associated with multiple reflections in an assembly of apparatus, which is seldom corrected because of the extra work involved and the practical difficulties to be overcome. Thus, if voltage reflection coefficients seen looking in opposite directions from a reference plane are Γ_1 and Γ_2, then the multiple reflection factor $|1 - \Gamma_1\Gamma_2|^{-2}$ is encountered in the formulae relating to attenuation. (See Chapter 2.) Now it is easy to measure the moduli $|\Gamma_1|$ and $|\Gamma_2|$ but it is far less easy to measure accurately the phase angles ϕ_1 and ϕ_2, especially when $|\Gamma_1|$ and $|\Gamma_2|$ are very small. Moreover, the reflections are usually stated merely as maximum modulus over a wide frequency range, over which the phase angles would vary considerably. We have

$$|1 - \Gamma_1\Gamma_2|^2 = 1 + |\Gamma_1|^2 \cdot |\Gamma_2|^2 - 2|\Gamma_1| \cdot |\Gamma_2| \cos(\phi_1 + \phi_2)$$

and the extreme limits with varying ϕ's are obtained when $\cos(\phi_1 + \phi_2) = \pm 1$. So if we define

$$a = 2|\Gamma_1| \cdot |\Gamma_2|$$

then the limits of the multiple reflection factor, neglecting a^4 and higher powers of a, are given by

$$1 + \tfrac{3}{4}a^2 \mp a\left(1 + \frac{a^2}{2}\right) \approx 1 \mp a$$

The resulting range in decibels is $M \approx \pm 4\cdot343a$. The basic assumption now made is that any value of $\phi_1 + \phi_2$ is equally likely. Bearing in mind the nature of a cosine function, the range of $x = \phi_1 + \phi_2$ needed to cover all values of the cosine once only is 0 to π. Then the *a priori* probability density distribution of the 'random' (actually unknown) variable x is uniform between 0 and π, of magnitude $1/\pi$, and zero outside these limits.

If one transforms from the variable x to the variable $y = a \cos x$ then one finds that y has a probability density distribution of U-shape between $y = \pm a$. Thus

$$g(y) = \frac{1}{\pi(a^2 - y^2)^{1/2}} \quad (y \text{ between } \pm a)$$

when

$$g(y) = 0 \quad (y \text{ outside } \pm a)$$

$$f(x) = 1/\pi \quad (x \text{ between 0 and } \pi)$$

$$f(x) = 0 \quad (x \text{ outside 0 to } \pi)$$

The mean square value of y is found to be $\frac{1}{2}a^2$ so the r.m.s. uncertainty is $0.707a$.

A slightly pessimistic but simple and effective approximation to this U-shaped distribution is that of two discrete values $\pm a$, each with a probability of $\frac{1}{2}$.

14.5 Combination of uncertainties and expression of the result

In general, it is desirable to know something about the resultant probability density distribution because of the ultimate wish to quote a range of uncertainty with a given confidence probability (confidence level) such as 0.95 or 0.99.

If, from the theory of the measurement process it is reasonably certain that the bounds of uncertainty of several assignable systematic errors are independent, then the theory of probability can be applied to obtain the resultant density distribution. The result is that two density functions $f_1(x)$ and $f_2(y)$ for random vrariables x and y lead to a probability density function $f_3(z)$ for $z = x + y$ obtained by *convolution* or *folding* in the form

$$f_3(z) = \int_{-\infty}^{\infty} f_1(x) f_2(z - x)\, dx \tag{14.5}$$

Several density functions are combined by successive folding. This integral is often difficult to evaluate, so, instead, one may make use of the important theorem in probability theory that, irrespective of the forms of the component probability density distributions, the mean square uncertainty (variance) of a linear function of independent random variables can be expressed in terms of the variances of all the random variables. Thus, if $z = c_1 x_1 + c_2 x_2 + \ldots$ is the linear function, then the resultant variance is

$$s^2 = c_1^2 s_1^2 + c_2^2 s_2^2 + \ldots \tag{14.6}$$

which is similar to eqn. 14.4. Use is also made of the fact that the resultant of several independent components of uncertainty, when they do not differ greatly in range, has a probability density distribution that is approximately normal, even though the components may be non-normal. It follows that the resultant distribution is often nearly normal and the limits for a given confidence level can be calculated from the resultant variance (given by eqn. 14.6) from a table of probability integral values.

In the extreme case where the range of one systematic uncertainty that is ascribed to a discrete-pair distribution (i.e. equal probability of $+a$ or $-a$) is dominant, the resultant range is simply $2a$ plus the range due to the resultant of all other contributions.

It is sometimes useful, and certainly instructive, to express distributions approximately as discrete-point distributions. (Expressed digitally, all results really have discrete-point distributions.) Two such distributions $h_1(x)$ and $h_2(y)$ can then be 'folded' to obtain the distribution of $z = x + y$, using the sum

$$h_3(z_m) = \sum_k h_1(x_k) h_2(z_m - x_k) \tag{14.7}$$

which corresponds to eqn. 14.5. The process is mechanical in its simplicity, and the result may be translated into a discrete-level histogram with $f(x)$. $\Delta x = h$, where Δx is the distance between adjacent pairs of points, $f(x)$ is the height of the histogram slice and h is the discrete probability height. The uncertainty $\pm u$ is determined for a given confidence level from the areas of the histogram inside and outside $\pm u$.

The results are expressed in the form $x \pm u$ (units) with a stated confidence probability (confidence level). The view taken here is that all uncertainties associated with systematic and random errors are represented by probability density distributions of some form (even those with only two discrete values of equal probability) and can be combined by folding to obtain an overall distribution. This does not prevent random and systematic components of uncertainty being expressed separately, which some applications of a measurement require. But it does enable a single overall uncertainty to be stated with a given confidence level, which calibrations by a standardising laboratory should require.

The mode of expression of the results of measurements depends on what use is to be made of the results. As much information as possible on the assessment of the uncertainty should be given, but in any case it is essential that the mode of expression is always indicated and is never vague.

14.6 Example of determination and expression of uncertainty

The example given to illustrate application of the foregoing methods is the measurement of relative attenuation between two settings of a rotary-vane microwave attenuator, using an i.f. substitution method with an i.f. piston attenuator as the local 'standard'. To take advantage

Table 14.1

P_c n	0·5 'probable error'	0·683 'standard error' (σ)	0·95 95% (approx. 2σ)	0·99 99%	0·997 3σ
3	0·82	1·3	4·3	9·9	–
4	0·76	1·2	3·2	5·8	9·5
5	0·74	1·2	2·8	4·6	6·5
6	0·73	1·1	2·6	4·0	5·4
7	0·72	1·1	2·5	3·7	4·9
8	0·71	1·1	2·4	3·5	4·5
9	0·71	1·1	2·3	3·4	4·2
10	0·70	1·1	2·3	3·2	4·0
12	0·70	1·1	2·2	3·1	3·8
15	0·69	1·0	2·1	3·0	3·6
20	0·69	1·0	2·1	2·9	3·4
∞	0·67	1·0	2·0	2·6	3·0

Table of 'Student-t' as a function of the number of measurements n with the confidence probability P_c as a parameter. The values are approximate, to two significant figures, which is considered adequate for most purposes.

of the opportunity afforded to enlarge on practical details of the calculation of uncertainties, two different cases will be treated: one where the i.f. substitution apparatus had been calibrated against the national standard and corrections were given and used, the other where no such calibration was made, so the systematic uncertainty arising from the imperfections in the i.f. substitution apparatus with its i.f. piston attenuator had to be estimated.

The nominal attenuation measured was 15 dB and the ten results actually obtained were, in order,

15·040	15·054
15·045	15·043
15·049	15·045
15·056	15·040
15·055	15·041

separated into two consecutive groups of five. The mean value $\bar{x}_1$ from eqn. 14.1 and the variance s_1^2 from eqn. 14.2 for the first group are

$$15 \cdot 0490 \, \text{dB and } 45 \cdot 5 \times 10^{-6} \, (\text{dB})^2$$

and $\bar{x}_2, s_2^2$ for the second group are

$$15 \cdot 0446 \, \text{dB and } 31 \cdot 3 \times 10^{-6} \, (\text{dB})^2$$

From these four statistics one can check for consistency, as described in Section 14.3. Thus $F_0 = s_1^2/s_2^2$, where $s_1 > s_2$ gives $F_0 = 45 \cdot 5/31 \cdot 3 = 1 \cdot 45$, which is much less than $6 \cdot 4$, so we conclude that the variance of each group is consistent with one population, the actual difference being due to sampling. Again, $\bar{x}_1 - \bar{x}_2 = 0 \cdot 0044 \, \text{dB}$, and as

$$(s_1^2 + s_2^2)^{1/2} = 0 \cdot 00876 \, \text{dB}$$

$$t_0 = 2 \cdot 24 \times 0 \cdot 0044/0 \cdot 00876 = 1 \cdot 125$$

which is much less than $2 \cdot 3$. So we conclude that the mean of each group is consistent with one population, and that the measurement is 'under statistical control'.

Use of all ten results in eqns. 14.1 and 14.2 gives the mean $\bar{x} = 15 \cdot 0468 \, \text{dB}$ and the r.m.s. deviation $s = 0 \cdot 0063 \, \text{dB}$. In eqn. 14.3, the Student-t for 10 results for a confidence level of $0 \cdot 99$ (which we choose) is found from the table to be $t = 3 \cdot 2$. Then eqn. 14.3 gives for the mean and the random component of uncertainty

$$15 \cdot 0468 \pm 0 \cdot 0064 \, \text{dB} \quad (\text{c.l.} = 0 \cdot 99)$$

and we write the random component of uncertainty in the form

$$U_r = 0 \cdot 0064 \, \text{dB} \quad (\text{c.l.} = 0 \cdot 99) \tag{14.8}$$

For the systematic components of uncertainty, the two cases mentioned at the beginning of this section are treated separately. First, assume the i.f. substitution apparatus has been calibrated against the national standard (e.g. via a transfer standard attenuator) and suppose the result has an overall uncertainty

$$U_{s1} = 0 \cdot 0081 \, \text{dB} \quad (\text{c.l.} = 0 \cdot 99) \tag{14.9}$$

with a normal distribution. When the i.f. substitution apparatus is then used to measure attenuation, any correction required by the calibration is applied, and this uncertainty U_{s1} enters as a systematic component (of the 2nd group mentioned in Section 14.3).

Other components of systematic uncertainty, not covered by the calibration, are for mismatch and leakage. The uncertainty on account of leakage for an attenuation measurement of 15 dB is quite negligible under all reasonable conditions, as can be seen from Fig. 2.9. For the uncertainty component from mismatch, we turn to eqn. 2.75 and take account of the fact that only the moduli of the various complex co-efficients are measured. Because the mismatch factor M_s is small, we can use the approximation

$$\log_{10}(1 + \epsilon) = 0.4343\,\epsilon$$

where $\epsilon \ll 1$. So we obtain

$$M_s = \pm 8.686\,\{|\Gamma_G|\,(|s_{11e}| + |s_{11b}|) + |\Gamma_L|\,(|s_{22e}| + |s_{22b}|)$$
$$+ |\Gamma_G\Gamma_L|\,(|s_{21b}|^2 + |s_{21e}|^2)\} \tag{14.10}$$

The notation is explained in Chapter 2. From the measurements we are given:

$$|\Gamma_G| = |\Gamma_L| = 0.005; \qquad |s_{11b}| = |s_{22b}| = 0.01;$$
$$|s_{11e}| = |s_{22e}| = 0.025;$$
$$|s_{12b}| = |s_{21b}| = 0.98 \text{ (equivalent to loss of } 0.175 \text{ dB);}$$
$$|s_{12e}| = |s_{21e}| = 0.1733 \text{ (equivalent to loss of } 0.175 + 15.0468 \text{ dB),}$$

and from eqn. 14.10 we obtain

$$M_s = \pm 0.0033\,\text{dB} \tag{14.11}$$

As argued in Section 14.4, the probability density distribution of this is likely to be U-shaped. The fact that the semi-range (0.0033 dB) of this is substantially less than the resultant $(U_r^2 + U_{s1}^2)^{1/2}$ of the U_r and U_{s1} ($= 0.0103$ dB), means that the combination of the mismatch contribution U_{s2} with the normally distributed U_r and U_{s1} produces a substantially normal distribution for the resultant. This, in turn, permits us to take the r.m.s. uncertainty of the U-shaped component and multiply it by the factor 2.6, as if the component were normally distributed, to obtain the 0.99 c.l. uncertainty component. Thus

$$U_{s2} = 2.6 \times 0.0033/\sqrt{2} = 0.00607\,\text{dB}$$

So for the overall uncertainty, according to eqn. 14.6 with the c's equal to unity:

$$U = (U_r^2 + U_{s1}^2 + U_{s2}^2)^{1/2}$$
$$= (0.00004096 + 0.00006561 + 0.00003684)^{1/2}$$
$$= 0.0120\,\text{dB} \quad (\text{c.l.} = 0.99)$$

Therefore, when the apparatus has been calibrated against the national standard, the result is expressed

$$15 \cdot 047 \pm 0 \cdot 012 \, \text{dB} \quad (\text{c.l.} = 0 \cdot 99) \tag{14.12}$$

where the figures have been suitably rounded. The distribution is substantially normal. For the *random component*, eqn. 14.8 gives

$$U_r = 0 \cdot 0064 \, \text{dB}$$

and the *systematic component* is $(U_{s1}^2 + U_{s2}^2)^{1/2} = 0 \cdot 010 \, \text{dB}$.

[If the mismatch uncertainty range had been substantially greater than the other components of uncertainty, then the U-shaped distribution of range $\pm a$ would best be regarded approximately as a 2-point discrete distribution, and the resultant overall uncertainty would be nearly

$$U = a + (U_r^2 + U_{s1}^2)^{1/2} \tag{14.13}$$

The overall distribution would no longer be normal if $a > (U_r^2 + U_{s1}^2)^{1/2}$ but would be double-humped.]

Now we take the second case where the i.f. substitution apparatus has not been calibrated and we have to estimate the systematic component of uncertainty associated with its use. The first component of uncertainty arises from imperfections in the rate and uniformity of the piston attenuator, as well as imperfection in displacement measurements, although the latter appears more in the random component than in the systematic components of uncertainty. This first component of systematic uncertainty is covered by the statement '$\pm 0 \cdot 01 \, \text{dB}$ per 10 dB' so that for 15 dB the range is $\pm 0 \cdot 015 \, \text{dB}$. In the absence of further information this is taken to have a uniform probability density distribution between the stated limits, i.e. a rectangular distribution. As with a U-shaped distribution, the combination of a rectangular distribution with a normal distribution is made approximately by a method that depends on whether the range of the non-normal distribution is greater than or less than the uncertainty range of the normal distribution. If the range is greater than that of the normal distribution then the combination is by simple addition as in eqn. 14.13. If the range is less than that of the normal distribution, then the combination is by quadrature, in accordance with eqn. 14.6 or in most simple cases more directly in terms of the U's with a given common c.l. under the assumption that they are all normal distributions determined from their actual r.m.s. values. This is exactly the same procedure that was followed in the work leading up to the result of eqn. 14.12. Such approximate methods of combining probability density distributions are nearly always adequate, and it is only when detailed information on the form of the resultant distribution is desired that use of eqns. 14.5 or 14.7 is necessary.

There could be a correction and a residual consequent uncertainty on account of the non-linearity of the frequency convertor and of the piston-attenuator scale at the higher levels of amplitude. But with an attenuation range of only 15 dB, the measurement would normally be done at a level below that at which non-linearity is appreciable and above that at which noise results in uncertainty. So we have $U_r = 0.0064$ dB (normal), a U-shaped distribution of 0.0033 dB, and a rectangular distribution of 0.015 dB. The first two can be combined to give a near-normal distribution $(U_r^2 + U_{s2}^2)^{1/2} = 0.0088$ where the value of $U_{s2} = 0.00607$ dB as before. Because 0.015 is greater than 0.0088 we use eqn. 14.13 and so $U = 0.015 + 0.0088 = 0.0238$. The result is therefore expressed

$$15.047 \pm 0.024 \text{ dB} \quad \text{(c.l. 0.99)} \tag{14.14}$$

The overall distribution will be near-normal, the random component is 0.0064 dB (normal) and the systematic component is 0.0183 dB (non-normal). Comparison of eqn. 14.14 with eqn. 14.12 shows an obvious advantage of calibrating the i.f. substitution apparatus through a transfer standard, namely the reduction of uncertainty. The true advantage is much greater, though, because of the danger of unforeseen systematic error arising from certain faults developing in the electronic parts of the apparatus. Such errors can be readily detected by frequent checking with the transfer standard if this is suitably stable.

14.7 References

1 DEMING, W.E. and BIRGE, R.T.: 'On the statistical theory of errors', *Rev. Mod. Phys.*, 1934, **6**, pp. 119–161

2 EISENHART, C.S.: 'Realistic evaluation of the precision and accuracy of instrument calibration systems', *J. Res. NBS*, 1963, **67C**, pp. 161–187

3 DIETRICH, C.F.: 'Uncertainty, calibration and probability' (Adam Hilger, 1973)

4 WILLIAMS, A., CAMPION, P.J. and BURNS, J.E.: 'Statement of results of experiments and their accuracy', *Nucl. Instrum. & Methods*, 1973, **112**, pp. 373–376

Chapter 15

Conclusions

15.1 Review of present-day techniques

When great accuracy is needed over an enormous dynamic range ($>$ 80 dB), the three main competitors are: a parallel i.f. substitution system (see Chapter 7), an a.f. substitution system of the type described by Clark (see Section 8.2) and a modulated sub-carrier system (see Section 8.3.).

Owing to low frequency noise from mixer diodes, an i.f. substitution system has a lower receiver noise factor than either of these a.f. substitution systems. On the other hand, a.f. circuits can be made more stable than i.f. circuits and an inductive voltage divider is a simpler and more accurate reference standard than an i.f. piston attenuator. The modulated sub-carrier system has the great advantage of requiring only one microwave oscillator, and its noise factor could be improved considerably by raising the modulation frequency from 1 kHz to 100 kHz (the highest frequency at which very accurate IVDs can be made at the present time).

When every possible precaution is taken, the accuracy obtainable with each of these systems varies from about ± 0.002 dB at low values of attenuation to about ± 0.02 dB at 80 dB (for a confidence level in each case of 99·7%). At low values of attenuation, the main sources of error are: imperfect matching at the insertion point, mixer non-linearity, connector uncertainties, zero drift and short term instrumental jitter. At very high values of attenuation, the main sources of error are: noise, leakage, spurious signals and accumulated errors in the reference standard. Very small changes in attenuation such as those produced by the type of transfer standard shown in Fig. 13.5 can, with extreme care, be measured to an accuracy of about ± 0.0002 dB, as in this case it is not necessary to make and break microwave connectors during the measurements. The accuracy figures stated in this paragraph have been more or less confirmed by the results achieved in three recent international attenuation intercomparisons that were sponsored by the Consultative Committee on Electricity (CCE) of the International Committee of Weights and Measures (CIPM). The countries that participated in these intercomparisons were: the USA, Canada Australia, Japan, Sweden, West Germany, Hungary, Italy, France and Britain. The full results obtained in these exercises have not yet been published, but they will appear in future reports from the organisers.

During the last 30 years, parallel i.f. substitution systems have been much more popular than series i.f. substitution systems because, in the latter case, greater gain stability is needed in the receiver and a highly linear i.f. amplifier is needed ahead of the piston attenuator. However, by careful design, a satisfactory solution can be achieved and a series i.f. substitution system, with reasonable accuracy and a dynamic range of 100 dB, has been commercially available for several years (see Table 7.1).

The power ratio methods of attenuation measurement have limited dynamic range (≈ 30 dB for single step measurements). If thin film thermoelectric power meters are used, special attention must be paid to their deviations from square law and it may be necessary to apply correction factors to the results.

For sheer stability, the power ratio attenuation measuring system described by Engen and Beatty in 1960 (see Section 5.3) has never been surpassed. Its success is attributed to two factors: the two bolometers in it are kept in an ultra-stable water bath and there are no modulation circuits in it to contribute to the jitter and drift.

When the greatest possible dynamic range is needed (> 100 dB), the best system to use is r.f. substitution (see Chapter 6), with a powerful source and a superheterodyne receiver. With this method of measurement, no error is caused by deviations from linearity in the mixer. To achieve high accuracy with a series r.f. substitution system, it is essential to have both an ultra-stable source and an ultra-stable receiver. A parallel r.f. substitution system requires a constant loss phase shifter. Both of these disadvantages can be overcome by using the configuration devised by Larson and Campbell (see Fig. 6.7) but the dynamic range is then greatly reduced. Unless extreme care is taken, the reference standard in an r.f. substitution system is not likely to be as accurate as the reference standard in an a.f. or i.f. substitution system (see Chapters 3 and 4). Fig. 6.6 shows that a sophisticated parallel r.f. substitution system is a complex expensive equipment. If a spectrum analyser is available, it can be used as the receiver (see Chapter 6), thereby saving a lot of extra work and expense

For the measurement of low values of attenuation (< 1 dB), very good results can be obtained with a small amount of equipment by using either the dual channel power ratio method (see Section 5.2) or a reflection coefficient method (see Chapter 9). A considerable improvement in accuracy can be achieved by using a zero-setting waveguide switch as shown in Fig. 9.3. When localised attenuation measurements are required on uniform lengths of waveguide, the fixed probe method of Somlo (see Section 12.7) is recommended.

The shuttle pulse method (see Section 12.4) is a popular one for measuring the attenuation in the long lengths of helix waveguide used in millimetre wave communication systems.

The Q measurement technique (see Section 12.5) is undoubtedly the best available method for determining extremely low values of attenuation, such as the loss in a precision waveguide connector.

The method using a Josephson junction in a superconducting ring (see Section 12.9) suffers from the great disadvantage of needing liquid helium. However, it does not depend on any precision machining and the only item needed to determine the results is a table of Bessel function roots, so it may eventually be adopted as a primary attenuation standard.

For workers who possess only a small amount of equipment, the techniques that should be considered are:

(a) the single power meter method described in Section 5.1
(b) Korewick's a.f. substitution system (see Section 8.1)
(c) the shorted slotted line method discussed in Section 12.6
(d) series r.f. substitution with a TFT power meter as the receiver, used in conjunction with the backing-off circuit shown in Fig. 6.5.

All of the methods discussed up to now in this chapter can be classed as fixed-frequency techniques. Many different swept-frequency techniques have been described in Chapter 10. These swept frequency techniques save a tremendous amount of time, resonances which can easily be missed using point-by-point methods are clearly revealed and the effects of adjustments made to the device under test can be seen immediately over a wide band.

The swept frequency attenuation measurement techniques that use d.c. amplification (see Fig. 10.9) are simple, but their dynamic range is limited by drift and $1/f$ noise to about 30 dB and their accuracy is typically ± 0.8 dB when coaxial components are used. In a fixed frequency system, extremely good matches can be obtained on each side of the insertion point and, with a reasonably well matched unknown (v.s.w.r. < 1.1), the mismatch uncertainty can be reduced to less than ± 0·002 dB (see Fig. 2.8). However, in a swept frequency measurement system, broad band matching is essential on each side of the insertion point and the mismatch error alone may exceed ± 0·1 dB.

Greater dynamic range can be obtained with a swept-frequency attenuation measuring system by chopping the microwave signal and using a ratiometer that contains tuned amplifiers (see Fig. 10.5). A high modulation frequency ($\approx$ 30 kHz) is recommended and a dynamic range of about 60 dB can then be achieved.

The calibration grid technique (see Figs. 10.7 and 10.8) eliminates the need to stay within the square law region of the diode detector and it gives a very worthwhile improvement in both accuracy and dynamic range.

The introduction of microwave network analysers in 1967 produced revolutionary changes in the microwave measurement field (see Chapter 11). These instruments measure both the real and imaginary parts of all four *s*-parameters of a 2-port device. The early network analysers were rapidly followed by computer-controlled network analysers in which the mismatch, tracking and directivity errors are eliminated by very ingenious calibration and computer-correction techniques (see Section 11.3). The most sophisticated computer-corrected network analysers, such as that shown in Fig. 11.3, give results which are almost as accurate as those obtained in microwave standards laboratories. Tasks which would take several days to perform using point-by-point techniques can be completed in a few minutes on an automatic network analyser. Fully comprehensive equipment of this type is extremely expensive (> £100,000) and, up to now, only organisations with a large amount of accurate measurement work to do have been able to justify the cost. Where they have been installed, nearly all of the microwave measurement work has been channelled through them and dozens of conventional microwave test benches have rapidly disappeared.

15.2 Future developments

There is a considerable market for microwave attenuator calibrators throughout the world, and, in the future, the firms that manufacture them will undoubtedly strive to develop models that are more accurate, easier to use, less expensive and more reliable than the present ones. Microprocessors are likely to be incorporated in them and it is expected that PROMs will be widely used to store correction factors for the imperfect components which are built into them. Following the success achieved by D.E. de Jersey (see Section 10.3 and Fig. 10.11), greater use is likely to be made in the future of single-sideband modulators. Further efforts will undoubtedly be made to reduce noise, drift, leakage, non-linearity errors, spurious signals and the errors in attenuation reference standards. In the future, switched-couplers (see Section 13.3) may be built into attenuator calibrators, so that they can be checked easily every day. This would eliminate the possibility of gross errors occurring due to hidden electronic faults. Tighter tolerances will be needed on the internal dimensions of waveguides and coaxial lines, and improved waveguide and coaxial connectors will be required, to reduce mismatch and connector repeatability errors. At the present time, these two sources of error are frequently the dominant ones when low values of attenuation are measured.

Several automated attenuation measurement systems have been described in this book (see Sections 5.4, 7.5, 8.3.5 and Chapters 10 and 11). In the future, even more effort is likely to be devoted to automation projects. Fully automatic attenuator calibrators give a higher work output, eliminate human errors and allow skilled workers to be moved from monotonous routine calibration tasks to more satisfying pursuits.

In the future, greater use may be made of the millimetre wave part of the spectrum,[1] so there is likely to be a growing need for millimetre wave attenuator calibrators, in fact, two papers on this very subject[2,3] appeared just before this final chapter was written.

In the national standards laboratories throughout the word, great efforts will be needed in the next few years to make sure that the national attenuation standards continue to be appreciably more accurate than the best commercially available attenuator calibrators and automatic network analysers.

This book shows that numerous different attenuation measurement techniques have been devised in the last 30 years. If history is any guide, many more new ones will be described in the next decade and it is hoped that this book will help readers to assess their significance.

15.3 References

1 MARDON, A.: 'Applications of millimetre waves', *in* BENSON, F.A. (Ed.): 'Millimetre and submillimetre waves' (Iliffe, 1969), chap. 23
2 WHITE, D.E.: 'A computer operated millimeter wave insertion and return loss measuring system' *CPEM Digest*, 1976, pp. 121–123
3 YAMAGUCHI, G.M., *et al.*: 'Design considerations and performance evaluation of a novel V-band network analyser/reflection test unit', *CPEM Digest*, 1976, pp. 124–127

Scattering parameters

With a microwave network having several ports (e.g. waveguide flanges or coaxial connector faces) in which voltages and currents (or electric and magnetic field strengths) are directly proportional to one another, it is possible to describe its properties by a set of linear relations between the entering and outgoing wave amplitudes at all of the ports. Thus, for the 2-port network shown in Fig. A1.1, a_1 and a_2 represent complex amplitudes for the waves entering ports 1 and 2, respectively; while b_1 and b_2 are complex amplitudes for the waves emerging from ports 1 and 2, respectively. In general, these complex amplitudes can be voltages, currents, electric field strengths or magnetic field strengths, but, throughout this book, the modulus of each complex amplitude is taken to be the *root-mean-square voltage* of a wave travelling in the stated direction. (The voltage measured by a voltmeter probe at any cross-section is the complex sum $a + b$ of the wave voltages travelling in each direction.) The relationships between the wave amplitudes in Fig. A1.1 are as follows:

$$b_1 = s_{11}a_1 + s_{12}a_2 \qquad (A1.1)$$

$$b_2 = s_{21}a_1 + s_{22}a_2 \qquad (A1.2)$$

The coefficients s_{11}, s_{12}, s_{21}, s_{22} are termed scattering parameters or scattering coefficients. These coefficients are complex quantities and are interpreted as follows:

When a generator is connected to port 1 and a non-reflecting (i.e. perfectly matched) termination is connected to port 2, $a_2 = 0$ and, from eqn. A1.1, we then get

$$s_{11} = \frac{b_1}{a_1} = \quad \text{voltage reflection coefficient looking into port 1 when port 2 is perfectly matched}$$

and, from eqn. A1.2, it is seen that

$$s_{21} = \frac{b_2}{a_1} = \text{voltage transmission coefficient from port 1 to port 2 when the latter port is perfectly matched}$$

When a generator is connected to port 2 and a perfectly matched termination is connected to port 1, $a_1 = 0$ and from eqn. A1.2

$$s_{22} = \frac{b_2}{a_2} = \text{voltage reflection coefficient looking into port 2 when port 1 is perfectly matched}$$

and, from eqn. A1.1,

$$s_{12} = \frac{b_1}{a_2} = \text{voltage transmission coefficient from port 2 to port 1 when port 1 is perfectly matched}$$

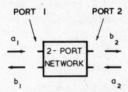

Fig. A1.1 Diagram showing the complex wave amplitudes entering and leaving a 2-port network

All these scattering parameters have to be relative to a given characteristic impedance and, in most problems, it is stipulated that the same nominal real characteristic impedance applies to each port.

In the case of a coaxial line, the characteristic impedance Z_0 is an unambiguous quantity that is defined as the input voltage between the inner and outer conductors divided by the input current when the line is either infinitely long or perfectly matched. At some point in such a line, at a distance x from the origin, let the complex forward wave amplitude be a_x. Then, according to the definition adopted for this book, $|a_x|$ denotes the r.m.s. value of the voltage between the inner and outer conductors at this point and so the forward power at a distance x from the origin is given, when Z_0 is real*, by

$$P_x = \frac{|a_x|^2}{Z_0} \qquad (A1.3)$$

---

*The effect of conductor resistance on the characteristic impedance is negligible at frequencies above 2 GHz

The more general case, where both forward and backward waves are present, is treated in Chapter 2 (see eqn. 2.5).

To give a precise meaning to voltage in a matched rectangular waveguide propagating the H_{01} mode, we define it at distance x as the root mean square voltage $|a_{x,wg}|$ across the narrow dimension l_b mid-way along the broad dimension l_a. The characteristic impedance of the waveguide $Z_{0,wg}$ is then defined in terms of this voltage and the power propagated along the waveguide. From standard texts on waveguide theory, it is found that

$$Z_{0,wg} = 2\frac{l_b}{l_a} \cdot \frac{\lambda_g}{\lambda} \sqrt{\frac{\mu}{\epsilon}} \tag{A1.4}$$

where λ_g is the guide wavelength, λ is the unbounded wavelength in the medium that fills the waveguide and μ and ϵ denote, respectively, the permeability and permittivity of this medium. With these definitions, the power at distance x, in an H_{01} wave that is travelling along a perfectly matched waveguide, is given by

$$P_{x,wg} = \frac{|a_{x,wg}|^2}{Z_{0,wg}} \tag{A1.5}$$

Finally, where reciprocity applies to a network, we have

$$s_{12} = s_{21} \tag{A1.6}$$

Signal flow graphs

Signal flow graphs enable calculations on microwave networks to be done in a rapid and systematic manner. They are particularly useful when several networks are connected in cascade.

In a signal flow graph,[1-7] complex wave amplitudes such as a_1, a_2, b_1 and b_2 in Fig. A1.1 are represented by points or nodes and the s parameters are represented by directed lines. Fig. A2.1 shows the signal flow graph for the two-port network given in Fig. A1.1. The arrow directions are from the independent variables (a_1 or a_2) to the dependent variables (b_1 or b_2). The value of a node is the sum of all signals entering it, each signal being the value of the node from which it comes multiplied by the path coefficient, e.g. in Fig. A2.1, $b_1 = s_{11}a_1 + s_{12}a_2$. This result is immediately seen to be in agreement with eqn. A1.1.

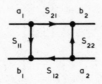

Fig. A2.1 Signal flow graph for a 2-port network

Fig. A2.2 shows the coaxial circuits, the signal flow graphs and the associated equations for:

(a) an imperfectly matched generator with a reflection coefficient Γ_g
(b) a termination with a reflection coefficient Γ_L
(c) a linear detector with a reflection coefficient Γ_d. In this case, K is a factor relating the incoming wave amplitude a to a meter reading M. The value of K will depend to some extent on Γ_d.

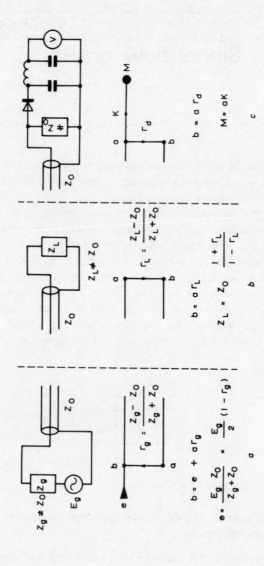

Fig. A2.2 Coaxial circuits, signal flow graphs and the associated equations

a For an imperfectly matched generator
b For an unmatched termination
c For an imperfectly matched linear detector

In Fig. A2.2a, E_g is the source e.m.f., i.e. the open-circuit voltage of the generator. In the corresponding signal flow graph, e is the voltage of the forward wave with a perfectly matched termination.

When several microwave networks are cascaded, the signal flow graphs of the individual networks can be joined together. The ratio between any two complex wave amplitudes (the typical network problem) can be written down either straight away, or after a little counting, by making use of a general formula known as Mason's non-touching loop rule. Before stating this formula, it is necessary to define paths and first, second and third order loops.

Path

In a signal flow graph, the value of a path is the product of all path coefficients encountered en route. A path must always follow the directions of the arrows and no node must be passed more than once. In Fig. A2.3, the value of one path from a_1 to b_1 is equal to s_{11} and the value of a second path from a_1 to b_1 is equal to $s_{21}\Gamma_L s_{12}$.

First-order loop

A first-order loop is a closed path which can be followed, always in the direction of the arrows, without passing any node more than once. Its value is the product of all path coefficients encountered en route. In Fig. A2.3, there are three first-order loops. Their values are: $\Gamma_g s_{11}$, $s_{22}\Gamma_L$ and $\Gamma_g s_{21}\Gamma_L s_{12}$.

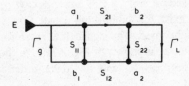

Fig. A2.3 Signal flow graph of a 2-port network between a generator and a load
This diagram is used to illustrate the meaning of: path, first-order loop, second-order loop etc.

Second-order loop

A second-order loop is the product of the values of any two first-order loops, which do not touch at any point. In Fig. A2.3, there is one second-order loop and its value is $\Gamma_g s_{11} s_{22}\Gamma_L$

Third-order loop

A third-order loop is the product of any three non-touching first-order loops and so on for higher-order loops.

Mason's non-touching loop rule

The ratio T of the complex wave amplitude at point Y to that at an independent point X is given by the following expression:

$$T = \frac{P_1(1 - \Sigma^1 L_1 + \Sigma^1 L_2 - \ldots) + P_2(1 - \Sigma^2 L_1 + \Sigma^2 L_2 - \ldots) + \ldots}{1 - \Sigma L_1 + \Sigma L_2 - \ldots}$$

(A2.1)

where

P_1 is one path from X to Y

P_2 is a different path from X to Y etc.

ΣL_1 is the sum of all 1st-order loops

ΣL_2 is the sum of all 2nd-order loops

ΣL_3 is the sum of all 3rd-order loops etc.

$\Sigma^1 L_1$ is the sum of all 1st-order loops not touching P_1

$\Sigma^1 L_2$ is the sum of all 2nd-order loops not touching P_1

$\Sigma^2 L_1$ is the sum of all 1st-order loops not touching P_2

$\Sigma^2 L_2$ is the sum of all 2nd-order loops not touching P_2

A proof of this rule has been given by C.S. Lorens.[2] It can be shown to be correct by choosing a fairly complicated microwave network and working out some particular complex wave amplitude ratio, firstly by straight-forward algebraic solution of the scattering equations and then by application of eqn. A2.1. The results are the same if Mason's rule has been applied properly, e.g. if no loops have been missed.

Various rules exist for simplifying signal flow graphs[5,7] and one example will now be given. The signal flow graph for a 2-port network followed by a detector is shown in Fig. A2.4a. In this diagram, there is one first order loop, $s_{22}\Gamma_d$ and no higher-order loops. Using eqn. A2.1, we find that

$$\frac{M}{a_1} = \frac{s_{21}K}{1 - s_{22}\Gamma_d}$$

(A2.2)

and

$$\frac{b_1}{a_1} = s_{11} + \frac{s_{21}\Gamma_d s_{12}}{1 - s_{22}\Gamma_d}$$

(A2.3)

Thus, the signal flow graph shown in Fig. A2.4a can be replaced by the much simpler one shown in Fig. A2.4b where

$$d_{21} = \frac{s_{12}K}{1 - s_{22}\Gamma_d}$$

(A2.4)

and

$$d_{11} = s_{11} + \frac{s_{21}\Gamma_d s_{12}}{1 - s_{22}\Gamma_d} \qquad (A2.5)$$

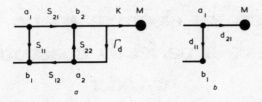

Fig. A2.4 Signal flow graph of a 2-port network followed by a detector
    a Full signal flow graph
    b Simplified version of signal flow graph shown in *a*

Here, and in Chapters 2, 9 and 11, symbols other than *s* are used to denote scattering parameters. This is unavoidable when different symbols have to be distinguished. However, they can be recognised as scattering parameters by the double suffix notation.

Evanescent electromagnetic fields inside a perfectly conducting cylinder

A3.1 Equations for field intensities

A cylindrical coordinate system r, ϕ, z is the obvious one to use so that the perfectly conducting internal surface of the cylinder is described simply by $r = a$. In this coordinate system, Maxwell's differential equations with a time dependence $e^{j\omega t}$ are expressed[*]:

$$j\omega\epsilon E_r = \frac{1}{r}\frac{\partial H_z}{\partial \phi} - \frac{1}{r}\frac{\partial (rH_\phi)}{\partial z} \tag{A3.1}$$

$$j\omega\epsilon E_\phi = \frac{\partial H_r}{\partial z} - \frac{\partial H_z}{\partial r} \tag{A3.2}$$

$$j\omega\epsilon E_z = \frac{1}{r}\frac{\partial (rH_\phi)}{\partial r} - \frac{1}{r}\cdot\frac{\partial H_r}{\partial \phi} \tag{A3.3}$$

$$-j\omega\mu H_r = \frac{1}{r}\frac{\partial E_z}{\partial \phi} - \frac{1}{r}\frac{\partial (rE_\phi)}{\partial z} \tag{A3.4}$$

$$-j\omega\mu H_\phi = \frac{\partial E_r}{\partial z} - \frac{\partial E_z}{\partial r} \tag{A3.5}$$

$$-j\omega\mu H_z = \frac{1}{r}\frac{\partial (rE_\phi)}{\partial r} - \frac{1}{r}\frac{\partial E_r}{\partial \phi} \tag{A3.6}$$

---

[*] Here, μ and ϵ are absolute permeability (H/m) and permittivity (F/m) for the macroscopic medium. One has $\mu = \mu_r\mu_0$ and $\epsilon = \epsilon_r\epsilon_0$ where μ_r, ϵ_r are the dimensionless relative values and μ_0, ϵ_0 are the absolute constants for a vacuum

The divergence equations do not have to be used explicitly here. The method of solution employed is that due to Bromwich in which the general solution is the sum of two partial solutions, one with $E_z = 0$ and the other with $H_z = 0$. It is sufficient for our purpose to consider only the case where $E_z = 0$.

Then eqn. A3.3 immediately gives

$$\frac{\partial(rH_\phi)}{\partial r} = \frac{\partial H_r}{\partial \phi}$$

which suggests that H_ϕ and H_r can be expressed as derivatives of a potential P, as follows:

$$rH_\phi = \frac{\partial P}{\partial \phi} \tag{A3.7a}$$

and

$$H_r = \frac{\partial P}{\partial r} \tag{A3.7b}$$

for then

$$\frac{\partial}{\partial r}(rH_\phi) = \frac{\partial^2 P}{\partial r \partial \phi} \quad \text{and} \quad \frac{\partial H_r}{\partial \phi} = \frac{\partial^2 P}{\partial \phi \partial r}$$

These are equal because the order of partial differentiation can be reversed without change. Then, with eqns. A3.7a and A3.7b in eqns. A3.4 and A3.5

$$-j\omega\mu \frac{\partial P}{\partial r} = -\frac{1}{r}\frac{\partial}{\partial z}(rE_\phi) \quad \text{and} \quad -\frac{j\omega\mu}{r}\frac{\partial P}{\partial \phi} = \frac{\partial E_r}{\partial z}$$

So

$$E_r = -\frac{j\omega\mu}{r} \int \frac{\partial P}{\partial \phi} dz$$

If, now,

$$P = \frac{\partial V}{\partial z}$$

then

$$\frac{\partial P}{\partial \phi} = \frac{\partial^2 V}{\partial \phi \partial z}$$

so that

$$\int \frac{\partial P}{\partial \phi} dz = \frac{\partial V}{\partial \phi}$$

Then

$$E_r = -\frac{j\omega\mu}{r}\frac{\partial V}{\partial \phi} \tag{A3.8a}$$

Similarly,

$$rE_\phi = j\omega\mu r \int \frac{\partial P}{\partial r} dz = j\omega\mu r \frac{\partial V}{\partial r}$$

or

$$E_\phi = j\omega\mu \frac{\partial V}{\partial r} \qquad (A3.8b)$$

Substitute for E_r, E_ϕ, H_r, H_ϕ from eqns. A3.8a, A3.8b, A3.7b, A3.7a and use in eqns. A3.1 and A3.2. Thus, in eqn. A3.1

$$\omega^2\mu\epsilon \frac{1}{r}\frac{\partial V}{\partial \phi} = \frac{1}{r}\frac{\partial H_z}{\partial \phi} - \frac{1}{r}\frac{\partial^3 V}{\partial z^2 \partial \phi}$$

or

$$\omega^2\mu\epsilon \frac{\partial V}{\partial \phi} = \frac{\partial}{\partial \phi}\left(H_z - \frac{\partial^2 V}{\partial z^2}\right) \qquad (A3.9)$$

In eqn. A3.2, we obtain in a similar way:

$$\omega^2\mu\epsilon \frac{\partial V}{\partial r} = \frac{\partial}{\partial r}\left(H_z - \frac{\partial^2 V}{\partial z^2}\right) \qquad (A3.10)$$

Integration of eqn. A3.9 choosing the constant to be zero, gives

$$H_z = \frac{\partial^2 V}{\partial z^2} + \omega^2\mu\epsilon V \qquad (A3.11)$$

If eqn. A3.11 is then substituted in eqn. A3.6, making use of eqns. A3.8a and A3.8b, there follows after some reduction:

$$\frac{1}{r}\frac{\partial}{\partial r}\left(r\frac{\partial V}{\partial r}\right) + \frac{1}{r^2}\frac{\partial^2 V}{\partial \phi^2} + \frac{\partial^2 V}{\partial z^2} + \frac{\omega^2}{v^2} V = 0 \qquad (A3.12)$$

where $v = (\mu\epsilon)^{-1/2}$ is the velocity of propagation in unrestricted space filled with the same dielectric as that which fills the space enclosed by the conducting cylinder. This partial differential equation for V can be solved by separation of variables by making the substitution

$$V = R(r)\cdot\Phi(\phi)\cdot Z(z)$$

On dividing eqn. A3.12 by V throughout,

$$\frac{1}{rR}\cdot\frac{d}{dr}\left(r\frac{dR}{dr}\right) + \frac{1}{r^2\Phi}\cdot\frac{d^2\Phi}{d\phi^2} + \frac{1}{Z}\cdot\frac{d^2 Z}{dz^2} + \frac{\omega^2}{v^2} = 0$$

The third term is a function of z alone while the first two terms together are functions only of r and ϕ. For the equation to be true for all r, ϕ, z, the two groups must be constants.

ω^2/v^2 is clearly constant. First we can let

$$\frac{1}{Z}\frac{d^2 Z}{dz^2} + \frac{\omega^2}{v^2} = k^2$$

so that

$$\frac{d^2Z}{dz^2} = \left(k^2 - \frac{\omega^2}{v^2}\right)Z = \gamma^2 Z \text{ (say)} \qquad (A3.13)$$

where

$$\boxed{\gamma^2 = k^2 - \frac{\omega^2}{v^2}} \qquad (A3.14)$$

The solution of eqn. A3.13 is $Z = A_1 e^{-\gamma z} + A_2 e^{+\gamma z}$. To simplify the immediate problem, let $A_2 = 0$ and $A_1 = 1$, so

$$Z = e^{-\gamma z} \qquad (A3.15)$$

Second, we have now

$$\frac{1}{rR} \cdot \frac{d}{dr}\left(r\frac{dR}{dr}\right) + \frac{1}{r^2\Phi}\frac{d^2\Phi}{d\phi^2} = -k^2 \qquad (A3.16)$$

and if this is multiplied by r^2

$$\frac{r}{R}\frac{d}{dr}\left(r\frac{dR}{dr}\right) + k^2 r^2 + \frac{1}{\Phi}\frac{d^2\Phi}{d\phi^2} = 0$$

The third term is a function of ϕ alone while the first two terms together are a function only of r. Let

$$\frac{1}{\Phi} \cdot \frac{d^2\Phi}{d\phi^2} = -n^2$$

so that

$$\frac{d^2\Phi}{d\phi^2} + n^2\Phi = 0 \qquad (A3.17)$$

which has the solution $\Phi = B_1 \cos(n\phi) + B_2 \sin(n\phi)$. We choose the zero of ϕ so that $B_2 = 0$, and taking $B_1 = 1$, we have

$$\Phi = \cos(n\phi) \qquad (A3.18)$$

For Φ to be single valued, n must be an *integer*. Now, on account of eqn. A3.17 we have in eqn. A3.16:

$$\frac{d^2R}{dr^2} + \frac{1}{r} \cdot \frac{dR}{dr} + \left(k^2 - \frac{n^2}{r^2}\right)R = 0 \qquad (A3.19)$$

where n is an integer on account of eqn. A3.18. This is Bessel's differential equation which has the general solution:

$$R = C_1 J_n(kr) + C_2 Y_n(kr)$$

Now $r = 0$ is included in the field in the present problem, so $Y_n(kr)$ must not be included in the solution because $Y_n(0)$ is infinite. Thus, $C_2 = 0$ and we take $C_1 = 1$, so

$$R = J_n(kr) \tag{A3.20}$$

Collecting eqns. A3.15, A3.18, A3.20, we finally obtain

$$\boxed{V = J_n(kr) \cdot \cos(n\phi) \cdot e^{-\gamma z}} \tag{A3.21}$$

To obtain the field strengths from V we have

$$E_r = -\frac{j\omega\mu}{r}\frac{\partial V}{\partial \phi} \tag{A3.8a}$$

$$E_\phi = j\omega\mu \frac{\partial V}{\partial r} \tag{A3.8b}$$

$$H_r = -\frac{j}{\omega\mu r}\frac{\partial}{\partial z}(rE_\phi) = \frac{\partial^2 V}{\partial z \partial r} \tag{A3.22}$$

from eqns. A3.4 and A3.8b.

$$H_\phi = \frac{j}{\omega\mu}\frac{\partial E_r}{\partial z} = \frac{\partial^2 V}{r\partial z\partial\phi} \tag{A3.23}$$

from eqns. A3.5 and A3.8a. Also,

$$H_z = \frac{\partial^2 V}{\partial z^2} + \frac{\omega^2}{v^2} V \tag{A3.11}$$

and from eqn. A3.14,

$$k^2 = \gamma^2 + \frac{\omega^2}{v^2} \tag{A3.24}$$

where k has yet to be determined. The resulting field intensities are:

$$\left\{ \begin{aligned} H_z &= k^2 J_n(kr) \cos(n\phi) \cdot e^{-\gamma z} & \text{(A3.25)} \\[2mm] H_r &= -\gamma k J_n'(kr) \cos(n\phi) \cdot e^{-\gamma z} & \text{(A3.26)} \\[2mm] H_\phi &= \frac{\gamma n}{r} J_n(kr) \sin(n\phi) \cdot e^{-\gamma z} & \text{(A3.27)} \\[2mm] E_r &= j\frac{\omega\mu n}{r} J_n(kr) \sin(n\phi) \cdot e^{-\gamma z} & \text{(A3.28)} \\[2mm] E_\phi &= j\omega\mu k J_n'(kr) \cos(n\phi) \cdot e^{-\gamma z} & \text{(A3.29)} \end{aligned} \right.$$

with an arbitrary multiplying constant for all five of the field amplitudes. Because $E_z = 0$, the different 'modes' labelled with values of the integer n are termed *H-modes*.

Because the electric field lines must meet the perfectly conducting surface of the cylinder perpendicularly, $E_\phi = 0$ at $r = a$. So in eqn. A3.29

$$\boxed{J'_n(ka) = 0} \tag{A3.30}$$

which determines the constant k. Now the Bessel functions J_n and their derivatives have an infinite number of zeros, so the values of $\rho = ka$ for which eqn. A3.30 holds have to be specified as ρ_{nm} where n is the integral order of the Bessel function and m is the order of the zero of $J'_n(\rho)$. From eqn. A3.14

$$\gamma = \sqrt{k^2 - \frac{\omega^2}{v^2}} = \sqrt{k^2 - \left(\frac{2\pi}{\lambda}\right)^2}$$

where λ is the free-space wavelength corresponding to v associated with the dielectric enclosed by the cylinder.

For an evanescent field, $k > 2\pi/\lambda$, i.e. λ is large compared with the diameter $2a$ of the cylinder. Then γ is *real* and is denoted by α. So

$$\alpha = k\sqrt{1 - \left(\frac{2\pi}{k\lambda}\right)^2} \tag{A3.31}$$

or if

$$k = \frac{2\pi}{\lambda_c}, \quad \text{then} \quad \alpha = \frac{2\pi}{\lambda_c}\sqrt{1 - (\lambda_c/\lambda)^2} \tag{A3.32}$$

where λ_c is the 'cut-off' wavelength in the dielectric of the waveguide.

For the solutions of eqns. A3.1 to A3.6 with $H_z = 0$, one obtains the following field intensities for the E-modes by a method analogous to that used to obtain eqns. A3.25 to A3.29.

$$\left\{\begin{array}{ll} E_z = k^2 J_n(kr) \cos(n\phi) e^{-\gamma z} & \text{(A3.33)} \\[2mm] E_r = -\gamma k J'_n(kr) \cos(n\phi) e^{-\gamma z} & \text{(A3.34)} \\[2mm] E_\phi = \dfrac{\gamma n}{r} J_n(kr) \sin(n\phi) \cdot e^{-\gamma z} & \text{(A3.35)} \\[2mm] H_r = -j\dfrac{\omega\epsilon n}{r} J_n(kr) \sin(n\phi) \cdot e^{-\gamma z} & \text{(A3.36)} \\[2mm] H_\phi = -j\omega\epsilon k J'_n(kr) \cdot \cos(n\phi) \cdot e^{-\gamma z} & \text{(A3.37)} \end{array}\right.$$

Because $E_\phi = 0$ at $r = a$, we now obtain

$$\boxed{J_n(ka) = 0}$$
(A3.38)

which determines k for the *E*-modes. Again, the number of zeros of J_n is infinite and the values of $\rho = ka$ for which eqn. A3.38 holds have to be specified by ρ_{nm} where m is the order of the zero of $J_n(\rho)$.

Also, for evanescent modes γ is real and is denoted by α, as before, and eqn. A3.31 or A3.32 still applies. We are concerned more with H-modes than with E-modes, except to note the possible presence of unwanted E-modes.

A3.2 Plotting the instantaneous lines of force

Plots of the lines of force of evanescent modes at a given instant are most useful in deciding the design of exciting and pick-up coils or loops, and in making a rough assessment of the types and relative magnitudes of unwanted modes which may result from any design of excitation system.

To calculate a line of force, as usually understood, in a given 2-dimensional plane such as (r, ϕ), one notices that the slope of a line of force at any point is given by the ratio of the corresponding field intensities at the point. Thus, for an electric field

$$\frac{1}{r}\frac{dr}{d\phi} = \frac{E_r}{E_\phi}; \quad \frac{dr}{dz} = \frac{E_r}{E_z} \quad \text{and} \quad r\frac{d\phi}{dz} = \frac{E_\phi}{E_z} \quad \text{(A3.39)}$$

These are often combined into a set of relations between differentials as follows:

$$\frac{dr}{E_r} = \frac{rd\phi}{E_\phi} = \frac{dz}{E_z} \quad \text{(A3.40)}$$

For actual calculations, the relations of eqn. A3.39 are of more direct use.

For the H_{1m} modes, we have from eqns. A3.28 and A3.29 and the first of eqn. A3.39:

$$\frac{E_\phi}{E_r} = r\frac{d\phi}{dr} = kr \cdot \frac{J_1'(kr)}{J_1(kr)} \cdot \frac{\cos\phi}{\sin\phi}$$

which, one should note, is independent of z. So

$$\int \frac{\sin\phi}{\cos\phi}d\phi = \int \frac{J_1'(kr)}{J_1(kr)}d(kr) = \int \frac{dJ_1(kr)}{J_1(kr)}$$

and

$$-\ln{(\cos{\phi})} = \ln{[J_1(kr)]} + \text{constant}$$

or

$$CJ_1(kr) = \sec{\phi}.$$

So the equation of a line of force for the electric field in any of the H_{1m} modes may be written

$$C \cdot J_1(\rho_{1m}r/a) = \sec{\phi} \qquad (A3.41)$$

Here, $k = \rho_{1m}/a$ where ρ_{1m} is the mth zero of $J_1'(\rho)$. The value of C for a given line of force is found from the value of r/a where the line of force cuts the line $\phi = 0$ for which $\sec{\phi} = 1$, and this value of C applies to the whole length of the given line of force.

To work out the equation of magnetic lines of force directly for H-modes can involve difficult integration and it is easier to note that eqns. A3.26 to A3.29 show that in the 'transverse' or r, ϕ plane the H-lines are everywhere perpendicular to the E-lines. This enables the H-lines to be plotted from a family of E-lines. Thus

$$\frac{H_\phi}{H_r} = -\frac{nJ_n(kr)\sin{(n\phi)}}{rkJ_n'(kr)\cos{(n\phi)}} = \left(\frac{rd\phi}{dr}\right)_H$$

while

$$\frac{E_r}{E_\phi} = \frac{nJ_n(kr)\sin{(n\phi)}}{rkJ_n'(kr)\cos{(n\phi)}} = \frac{-1}{(rd\phi/dr)_E}$$

The last relation shows that an H-line is perpendicular to the E-line at a given point (i.e. where they cross) in any transverse plane. Likewise, in the r, z plane difficulties with integration arise, and it is easier to plot the 'longitudinal' field in the ϕ, z cylindrically curved plane, usually at $r = a$. Thus from eqns. A3.25 and A3.27 for the H_{1m} modes (with $\gamma = \alpha$, of course)

$$\frac{H_z}{H_\phi} = \frac{k^2 r}{\alpha}\frac{\cos{\phi}}{\sin{\phi}} = \frac{dz}{rd\phi}$$

so that

$$z = \frac{(kr)^2}{\alpha} \int \frac{\cos{\phi}}{\sin{\phi}} d\phi = \frac{(kr)^2}{\alpha} \ln{(\sin{\phi})} + z_0$$

or

$$z - z_0 = -\frac{(kr)^2}{\alpha} \ln{(\text{cosec }\phi)} \qquad (A3.42)$$

where r is constant and z_0 is an arbitrary constant. At $r = a$ (near the inside surface of the cylinder), $ka = \rho$ and $\alpha \simeq k$ when λ is large compared with $2a$, and we have

$$\frac{z - z_0}{a} = -\rho_{1m} \ln{(\text{cosec }\phi)} \qquad (A3.43)$$

Examples of lines of force are shown in Figs. A3.1, A3.2 and A3.3. Fig. A3.1 shows one quadrant of a transverse section of an H_{11} mode where the continuous lines denote the electric field derived from eqn. A3.41 with $\rho_{11} = 1\cdot841$ and the broken lines denote the lines of magnetic force, drawn orthogonal to the lines of electric force. Fig. A3.2 shows one quadrant of a transverse section of an H_{12} mode, derived from eqn. A3.41 with $\rho_{12} = 5\cdot331$. The broken lines denote the magnetic field as before.

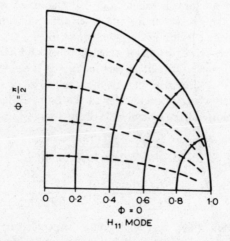

Fig. A3.1 One quadrant of a transverse section in a cylindrical waveguide
Magnetic lines of force of H_{11} mode are shown by broken lines

Fig. A3.3 shows the lines of magnetic force near $r = a$ in the developed half cylinder from $\phi = 0$ to π and along z. The family is drawn so that the density of the lines is roughly proportional to the intensity at any point. It is derived from eqn. A3.43 with $\rho_{11} = 1\cdot841$ for the H_{11} mode shown in Fig. A3.3a and with $\rho_{12} = 5\cdot331$ for the H_{12} mode shown in Fig. A3.3b. The latter shows a higher rate of attenuation than (a), as one would expect from the values of ρ. The transverse sections of the lines of force for all values of z have the same form as Fig. A3.1 for the H_{11} mode and the same form as Fig. A3.2 for the H_{12} mode, but the *intensity* is attenuated according to $e^{-\rho z/a}$ as z increases.

The magnetic field in the important $z-r$ plane through $\phi = 0$ can be sketched by approximate means and the general forms are shown in Fig. A3.4 a and b. These figures all show clearly that evanescent modes are no more than fringe fields governed by the forced field in the plane of excitation.

A3.3 Characteristic impedance

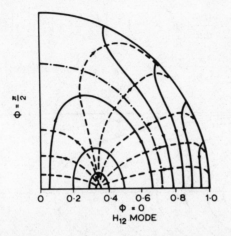

Fig. A3.2 One quadrant of a transverse section in a cylindrical waveguide
Magnetic lines of force of H_{12} mode are shown by broken lines

Considerations are confined to the H_{11} mode and we start from eqns. A3.26 to A3.29 with $n = 1$ and $\gamma = \alpha$ and where k is determined by the first root of $J_1'(ka) = 0$. In order to express all amplitudes in terms of the transverse electric field intensity E_0 at the centre of the cylinder, multiply eqns. A3.26 to A3.29 by $-2jE_0/(\omega\mu k)$. Then

$$E_r = E_0 \cdot \frac{2J_1(kr)}{kr} \cdot \sin\phi\, e^{-\alpha z} \tag{A3.44}$$

$$E_\phi = E_0 \cdot 2J_1'(kr) \cdot \cos\phi\, e^{-\alpha z} \tag{A3.45}$$

$$H_r = -\frac{j\alpha}{\omega\mu}(-E_0) \cdot 2J_1'(kr) \cdot \cos\phi\, e^{-\alpha z} \tag{A3.46}$$

$$H_\phi = -\frac{j\alpha}{\omega\mu} \cdot (E_0) \cdot \frac{2J_1(kr)}{kr} \cdot \sin\phi\, e^{-\alpha z} \tag{A3.47}$$

In these equations, the amplitudes E and H are taken to be r.m.s. values. In eqns. A3.46 and A3.47, the $-j$ factor shows that the oscillating magnetic field is in quadrature with and lagging the electric field. As a result there is no power transmitted in the simple evanescent mode. The energy flux or power calculated by the Poynting vector is pure imaginary. Thus

$$\Pi_z = \int_0^{2\pi} d\phi \int_0^a (E_r H_\phi^* - E_\phi H_r^*)\, r dr$$

$$= \frac{j\alpha}{\omega\mu} 4\pi E_0^2 \, e^{-2\alpha z} \int_0^a \left\{ \frac{J_1^2(kr)}{(kr)^2} + J_1'^2(kr) \right\} r dr$$

$$= \frac{j\alpha}{\omega\mu} \cdot \frac{4\pi E_0^2}{k^2} e^{-2\alpha z} \int_0^\rho \left\{ \frac{J_1^2(x)}{x} + x J_1'^2(x) \right\} dx$$

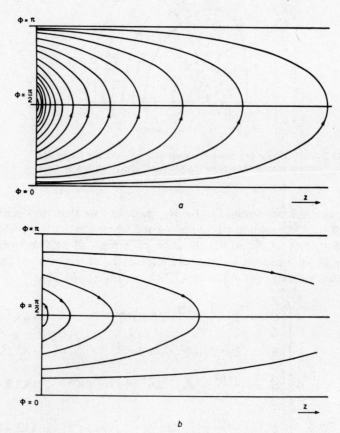

Fig. A3.3 Lines of magnetic force in the developed half cylinder $\phi = 0$ to π at radius a

a H_{11} mode
b H_{12} mode

where $x = kr$, $\rho = ka$ and Π_z is a complex Poynting vector. The real power is $P_z = \text{Re}\,\Pi_z$. By use of the standard recurrence formulae for Bessel functions and the Lommel integral, the integral can be evaluated with the result

$$\Pi_z = \frac{j\alpha}{\omega\mu} \cdot \frac{4\pi E_0^2}{k^2} \cdot e^{-2\alpha z} \cdot \frac{\rho^2}{2}\left(1 - \frac{1}{\rho^2}\right)J_1^2(\rho) \qquad (A3.48)$$

when $J_1'(\rho) = 0$. (The notation $J_1^2(\rho)$ means $[J_1(\rho)]^2$.) For the H_{11} mode, we have $\rho = 1\cdot841$ and $k = 1\cdot841/a$ so that

$$\Pi_z = j\frac{\alpha}{\omega\mu} \cdot 1\cdot5 E_0^2 a^2 \cdot e^{-2\alpha z} \qquad (A3.49)$$

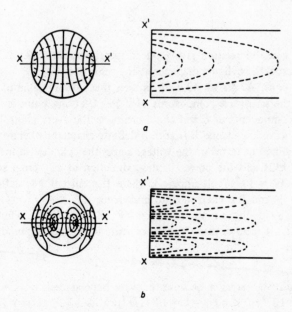

Fig. A3.4 Magnetic lines of force in the diametral plane through $\phi = 0$
Magnetic lines of force are shown by broken lines. Electric lines of force in the transverse plane are shown by unbroken lines
a H_{11} mode
b H_{12} mode

From the point of view of coupling the electromagnetic field in the cylinder to lumped circuit elements, it is useful to express E_0 in terms of the voltage across the diameter of the cylinder from $\phi = \pi/2$ to $3\pi/2$. Assume that the origin lies in plane S (in Fig. 3.19) at which $z = 0$. Then, this voltage is

$$V_0 = 2 \int_0^a E_r dr = \frac{4E_0}{k} \int_0^\rho \frac{J_1(kr)}{kr} d(kr)$$

or

$$V_0 = \frac{4E_0}{k} \left\{ \int_0^\rho J_0(x)dx - J_1(\rho) \right\} \qquad (A3.50)$$

$(\sin \phi = 1)$

where $\rho = ka$. For $ka = 1\cdot841$, then

$$V_0 = 1\cdot744\, E_0 a \qquad (A3.51)$$

$$V_0^2 = 3\cdot05\, E_0^2 a^2 \qquad (A3.52)$$

In eqn. A3.49,

$$\Pi_z = j\frac{\alpha}{\omega\mu} \cdot \frac{V_0^2}{2\cdot03} e^{-2\alpha z} \qquad (A3.53)$$

which is a useful relation giving the imaginary power in terms of the voltage across the cylinder diameter at any value of z.

From eqns. A3.44 to A3.47 it is seen that, at any value of z, the ratio of the transverse components of E and H at any point is $j\omega\mu/\alpha$, and the components of E and H are orthogonal at every point. We say that the 'wave impedance' is $j\omega\mu/\alpha$. A definite characteristic impedance can be defined in terms of the voltage across the cylinder (in the direction $\phi = \pi/2$) and the power in the z-direction at any cross section. We have $\Pi_z = VI^*$ (volt-amperes) where the current I is defined in terms of V and the characteristic impedance Z_0. Now $I^* = (V/Z_0)^*$ $= V^*/Z_0^*$ by definition, so that $\Pi_z = VV^*/Z_0^* = |V|^2/Z_0^*$. Thus from eqn. A3.53 it follows that the characteristic impedance is pure imaginary and is

$$Z_0 = jX_0 = j2\cdot03\, \omega\mu/\alpha \qquad (A3.54)$$

It is a positive reactance because the wave impedance is positive. Now $\mu = 4\pi \times 10^{-7}$ H/m and $\alpha \approx k = 1\cdot841/a$ (provided $\lambda \gg 2a$) so

$$jX_0 = j1\cdot1\ \omega\mu a (\text{ohms}) \qquad (A3.55)$$

or

$$jX_0 = j0\cdot087 fa (\text{ohms}) \qquad (A3.56)$$

where f is in MHz and a is in cm.

A3.4 Application of microwave network theory

Remembering the results of the last section regarding wave impedance, let E_0 and H_0 be taken at the centre of the cross-section of the cylinder *at the source end S* (refer to Fig. 3.19). Then at the load L we have

$E_0 e^{-\alpha l}$ and $H_0 e^{-\alpha l}$. If we now define Γ_S and Γ_L as the 'voltage' reflection coefficients relative to the imaginary characteristic impedance jX_0, or the 'electric field' reflection coefficients relative to the characteristic wave-impedance $j\omega\mu/\alpha$, then, because of the multiple reflections between the two planes S and L, the standard microwave network theory gives for the resultant values of E and H at the centre of the cylinder at L:

$$E_L = \frac{E_0 e^{-\alpha l} (1 + \Gamma_L)}{1 - \Gamma_S \Gamma_L e^{-2\alpha l}} ; \quad H_L = \frac{H_0 e^{-\alpha l} (1 - \Gamma_L)}{1 - \Gamma_S \Gamma_L e^{-2\alpha l}} \quad \text{(A3.57)}$$

The factor $(1 + \Gamma_L)$ in E_L accounts for the superposition of incident and reflected fields at L; likewise the factor $(1 - \Gamma_L)$ in H_L does the same for the magnetic field, bearing in mind that $-\Gamma_L$ is the magnetic field reflection coefficient. The denominators account for the effect of multiple reflections between S and L. The term 'reflection' is used here in a purely formal way because the phase velocity in an evanescent mode approaches an infinite value. But for the very small losss in the cylinder walls, it would be infinite. The system is more like a superposition of oscillating fields rather than a superposition of multiply reflected waves. Nevertheless, formal application of the reflected wave theory, but with an imaginary characteristic impedance, does lead to correct results. The power at L is $P_L = \text{Re}\,(V_L I_L^*)$. (A3.58)

Because V_L is linearly related to E_L and I_L is linearly related to H_L, so that Γ_L is the same for V or E and I or H, then eqns. A3.57 show that we may write for eqn. A3.58:

$$P_L = \frac{\text{Re}\,[V_0 I_0^* (1 + \Gamma_L)(1 - \Gamma_L^*)]\, e^{-2\alpha l}}{|1 - \Gamma_S \Gamma_L\, e^{-2\alpha l}|^2} \quad \text{(A3.59)}$$

Now $I_0^* = V_0^*/Z_0^*$ and $Z_0 = jX_0$ so:

$$\text{Re}\,[V_0 I_0^* (1 + \Gamma_L)(1 - \Gamma_L^*)] = \tfrac{1}{2}\,[V_0 I_0^* (1 + \Gamma_L)(1 - \Gamma_L^*)$$
$$+ V_0^* I_0 (1 + \Gamma_L^*)(1 - \Gamma_L)]$$

$$= \frac{|V_0|^2}{2} \left\{ \frac{(1 + \Gamma_L)(1 - \Gamma_L^*)}{Z_0^*} + \frac{(1 + \Gamma_L^*)(1 - \Gamma_L)}{Z_0} \right\}$$

$$= \frac{|V_0|^2}{2} \left\{ \frac{j(1 + \Gamma_L - \Gamma_L^* - \Gamma_L \Gamma_L^*)}{X_0} - \frac{j(1 + \Gamma_L^* - \Gamma_L - \Gamma_L \Gamma_L^*)}{X_0} \right\}$$

$$= \frac{|V_0|^2 \, j (\Gamma_L - \Gamma_L^*)}{X_0}$$

and, returning to eqn. A3.59, we finally get

$$P_L = \frac{|V_0|^2 \, j (\Gamma_L - \Gamma_L^*) \, e^{-2\alpha l}}{X_0 \, |1 - \Gamma_S \Gamma_L \, e^{-2\alpha l}|^2} \tag{A3.60}$$

This important relation is used in the theory of the piston attenuator in Chapter 3.

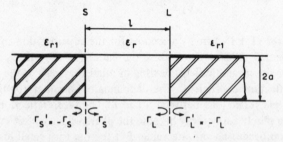

Fig. A3.5 Microwave piston attenuator
  The solid dielectric end-parts of the cylinder support a propagating mode, and the variable-length centre part supports an evanescent mode

A3.5 Application to the theory of microwave piston attenuators

Refer to Fig. A3.5. The radius a of the conducting cylinder and the relative permittivity ϵ_{r1} of the solid dielectric are chosen so that at the microwave frequency the H_{11} electromagnetic field is evanescent in the air-filled part of the cylinder but is propagated in the solid dielectric-filled parts of the cylinder. The fields at each boundary between air and solid dielectric are related by equality of the electric and magnetic field intensities, respectively, parallel to the boundary and immediately on each side of the boundary. The longitudinal component of magnetic field intensity is unchanged across the boundary because the permeability is the same on each side. It follows that the lines of force in a transverse section have the same form both in the solid dielectric and in the air-filled parts of the cylinder. This means that at the boundary S a pure H_{11} evanescent mode is set up in the air-filled cylinder by the H_{11} propagated mode in the solid dielectric-filled cylinder, provided no propagated E_{01} mode is present in the last mentioned part of the cylinder. In the solid dielectric let the unbounded velocity of propagation be $v_1 = (\mu \epsilon_1)^{-1/2}$ where $\epsilon_1 = \epsilon_{r1} \epsilon_0$, and let the unbounded wavelength be λ_1. In air, let the free-space velocity of propagation be $v = (\mu\epsilon)^{-1/2}$ where $\epsilon = \epsilon_r\epsilon_0$ and $\epsilon_r \sim 1.0006$ under normal conditions. Then, if ϵ_{r1} is large enough, it is possible to have a propagated H_{11} mode in the solid dielectric for which

$$\beta = \frac{2\pi}{\lambda_g} = \frac{2\pi}{\lambda_1}\sqrt{1 - \left(\frac{\lambda_1}{\lambda_c}\right)^2} \tag{A3.61}$$

and an evanescent H_{11} mode in the air dielectric for which

$$\alpha = \frac{2\pi}{\lambda_c}\sqrt{1 - \left(\frac{\lambda_c}{\lambda}\right)^2} \tag{A3.62}$$

where $\lambda > \lambda_c > \lambda_1$. These expressions follow from the development in Section A3.1.

In the solid dielectric the wave impedance (transverse E/H ratio) is real and is

$$\zeta_{01} = \frac{\omega\mu}{\beta} \tag{A3.63}$$

and in the air dielectric the transverse E/H ratio is imaginary and is expressed

$$\zeta_0 = j\frac{\omega\mu}{\alpha} \tag{A3.64}$$

as can be deduced from eqns. A3.26 to A3.29. From eqns. A3.63 and A3.64, we have with eqns. A3.61 and A3.62

$$j\frac{\zeta_{01}}{\zeta_0} = \frac{\alpha}{\beta} = \frac{\lambda_1}{\lambda_c}\sqrt{\frac{1 - (\lambda_c/\lambda)^2}{1 - (\lambda_1/\lambda_c)^2}} = r_z \text{ (say)} \tag{A3.65}$$

Then the voltage reflection coefficient Γ at each interface, relative to the imaginary wave impedance of the air-filled cylinder is

$$\Gamma = \frac{\zeta_{01} - \zeta_0}{\zeta_{01} + \zeta_0} = \frac{r_z - j}{r_z + j} = \frac{r_z^2 - 1 - 2jr_z}{r_z^2 + 1} \tag{A3.66}$$

If Γ, which stands for Γ_a or Γ_b in Fig. A3.5, is expressed in the form $|\Gamma|e^{j\psi}$, then it is found from eqn. A3.66 that $|\Gamma| = 1$ whatever the value of r_z, and

$$\tan\psi = \frac{2r_z}{1 - r_z^2} \tag{A3.67}$$

In a microwave piston attenuator, it is more useful to use voltage-reflection coefficients Γ' relative to the real wave impedance of the solid dielectric filled parts of the cylinder. We have, simply,

$$\Gamma' = \frac{\zeta_0 - \zeta_{01}}{\zeta_0 + \zeta_{01}} = -\Gamma \qquad (A3.68)$$

on using the first equality in eqn. A3.66. So

$$\Gamma' = -e^{j\psi} \qquad (A3.69)$$

where ψ is given by eqn. A3.67.

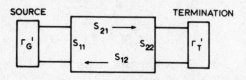

Fig. A3.6 2-port representation of microwave piston attenuator
 Scattering and reflection coefficients are expressed relative to the real
 characteristic wave impedance of the solid-dielectric-filled end sections

The air-spaced length of cylinder and the interfaces are best represented by a two-port element with scattering coefficients $s_{11} = s_{22}$
and $s_{21} = s_{12}$, the equalities arising from the symmetry of the cylinder.
The power absorbed in a load of reflection coefficient Γ'_T, when the
source has a reflection coefficient Γ'_G is expressed:

$$P_T = \text{const.} \times \frac{|s_{21}|^2 (1 - |\Gamma'_T|^2)}{|(1 - \Gamma'_G s_{11})(1 - \Gamma'_T s_{22}) - \Gamma'_G \Gamma'_T s_{21} s_{12}|^2} \qquad (A3.70)$$

where all reflection and scattering coefficients are relative to the real
ζ_{01}. Refer to Fig. A3.6.

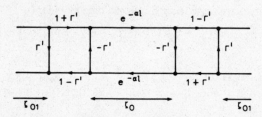

Fig. A3.7 Flow-graph of microwave piston attenuator
 Coefficients are expressed relative to the real characteristic wave impedance ζ_{01}

It remains to express $s_{11} = s_{22}$ and $s_{21} = s_{12}$ in terms of the coefficients $-\Gamma_S = -\Gamma_L = \Gamma' = -e^{j\psi}$ as given by eqn. A3.69. This can be
done by an application of flow-graph analysis, as shown in Fig. A3.7.

Each interface is represented by a simple 2-port where, taking that one on the left, Γ' is the v.r.c. of the interface viewed from the left and $1 + \Gamma'$ is the voltage transmission coefficient from left to right across the interface. This last coefficient arises from the boundary condition that the transverse electric field intensity parallel to the boundary must be the same on each side of the boundary, close to it, and the resultant transverse electric field on the left of the boundary is $E_+(1 + \Gamma')$ where E_+ is the intensity of the incident wave near the boundary. Looking to the left from the right of the boundary, then relative to the wave impedance ζ_0 of the air-filled cylinder, the reflection coefficent is $-\Gamma'$ and the transmission coefficient is $1 - \Gamma'$, by the same argument as before.

The voltage reflection coefficient of the unloaded network represented in Fig. A3.7, from either end, works out as

$$s_{11} = s_{22} = \frac{\Gamma'(1 - e^{-2\alpha l})}{1 - (\Gamma')^2 \, e^{-2\alpha l}} \tag{A3.71}$$

and the voltage transmission coefficient of the unloaded network, in either direction, is found to be

$$s_{21} = s_{12} = \frac{\{1 - (\Gamma')^2\} e^{-\alpha l}}{1 - (\Gamma')^2 \, e^{-2\alpha l}} \tag{A3.72}$$

If we now apply eqn. A3.69, we have

$$s_{11} = s_{22} = \frac{-e^{j\psi}(1 - e^{-2\alpha l})}{1 - e^{2j\psi} \, e^{-2\alpha l}} \tag{A3.73}$$

$$s_{21} = s_{12} = \frac{(1 - e^{2j\psi}) e^{-\alpha l}}{1 - e^{2j\psi} e^{-2\alpha l}} \tag{A3.74}$$

By substituting eqns. A3.73 and A3.74 for the s-parameters in eqn. A3.70, the nonlinearity of the piston attenuator resulting from interaction can be determined.

As an example, let us take $\Gamma'_T = 0$ which means that the load is matched to the dielectric-filled cylinder with its H_{11} transmission mode. This results in a reasonably tight coupling. Then in eqn. A3.70

$$P_T = \text{const.} \times \frac{|s_{21}|^2}{|1 - \Gamma'_G s_{11}|^2} \tag{A3.75}$$

$$= \text{const.} \times \frac{|1 - e^{2j\psi}|^2 \, e^{-2\alpha l}}{|1 - e^{2j\psi} e^{-2\alpha l} + \Gamma'_G e^{j\psi}(1 - e^{-2\alpha l})|^2}$$

If, now, Γ_G' can be made equal to $-e^{j\psi}$, the denominator becomes $|1 - e^{2j\psi}|^2$ and so

$$P_T = \text{const.} \times e^{-2\alpha l} \tag{A3.76}$$

where the non-linearity is absent. Such a value of Γ_G' as $-e^{j\psi}$ can be achieved by a short-circuit disc a chosen distance from the interface between solid and air dielectrics. The excitation arrangement would have to be such that it did not upset the value of Γ_G'. A loop with stabilised current would be suitable if the practical difficulties of achieving this could be overcome.

Another case, easier to realise, is with both Γ_T' and Γ_G' equal to zero, i.e. with matched source and load in the solid dielectric filled waveguide. Then, from eqn. A3.70

$$P_T = \text{const.} \times |s_{21}|^2$$

$$= \text{const.} \times \frac{|1 - e^{2j\psi}|^2 \, e^{-2\alpha l}}{|1 - e^{2j\psi} \, e^{-2\alpha l}|^2} \qquad \text{from (A3.74)}$$

and we find

$$P_T = \text{const.} \times \frac{1 - \cos 2\psi}{\cosh 2\alpha l - \cos 2\psi} \tag{A3.77}$$

Now ψ is given by eqn. A3.67 and a likely range of r_z is $1 \cdot 0$ to $2 \cdot 414$ giving ψ the range $\pi/2$ to $3\pi/4$ or $\cos 2\psi$ from -1 to 0. When $r_z = 2 \cdot 414$, we find from eqn. A3.77

$$P_L = \text{const.} \times \frac{2e^{-2\alpha l}}{1 + e^{-4\alpha l}} \tag{A3.78}$$

When $r_z = 1 \cdot 0$, we find

$$P_L = \text{const.} \times \frac{4e^{-2\alpha l}}{1 + 2e^{-2\alpha l} + e^{-4\alpha l}} \tag{A3.79}$$

and it is clear from the denominators in eqns. A3.78 and A3.79 that $r_z = 2 \cdot 414$ gives a much shorter range of non-linearity (for a given uncertainty) than does $r_z = 1 \cdot 0$.

For example, if the diameter of the bore, $2a = 10 \cdot 7\,\text{mm}$, then $(\lambda_c)H_{11} = 18 \cdot 26\,\text{mm}$ and $(\lambda_c)E_{01} = 13 \cdot 98\,\text{mm}$ (see the note at the end of this Appendix). The unbounded wavelength λ_1 in the solid dielectric must be greater than $(\lambda_c)E_{01}$ to ensure high attenuation of the E_{01} mode in the filled regions of the waveguide. The dielectric rods then act as mode filters. Let the frequency of the applied signal be exactly 9 GHz. First, let $\epsilon_r = 5$; then $\lambda_1 = 14 \cdot 9\,\text{mm}$, $\lambda_1/\lambda_c = 0 \cdot 816$ and $\lambda_c/\lambda = 0 \cdot 548$. From eqns. A3.65 and A3.67, we now get $r_z = 1 \cdot 180$ and $\tan\psi = -6 \cdot 001$. From this, we find that $\psi = 99 \cdot 5°$; so

$\cos(2\psi) = -0.946$ and, according to eqn. A3.77, the range of non-linearity is large. Secondly, let $\epsilon_r = 3.75$, then $\lambda_1 = 17.2$ mm, $\lambda_1/\lambda_c = 0.942$ and $\lambda_c/\lambda = 0.548$ as before. It now turns out that $r_z = 2.35$ and $\tan\psi = -1.038$. From this, we find that $\psi = 133.9°$; so $\cos(2\psi) = -0.038$, which is close to zero. Now, the range of non-linearity is near the minimum value. These examples show that the length of the appreciably non-linear region on the attenuation scale is strongly affected by the choice of relative permittivity of the solid dielectric in the propagated mode sections of the cylindrical waveguide.

Note

In this appendix, the cut-off wavelength λ_c, is defined as the unbounded wavelength in the solid dielectric associated with the cut-off frequency. This differs from the definition often adopted where λ_c is the free-space (vacuum) unbounded wavelength associated with the cut-off frequency, but the former is considered to be more directly applicable to the problems discussed in Section A3.5.

Combined effect of end vane misalignment and insufficient central vane attenuation on the performance of a rotary vane attenuator

Assume that the end vanes are correctly mounted in the two end sections which are misaligned by $+\epsilon$ and $-\epsilon$ relative to the datum (see Fig. A4.1). (With a total misalignment between the end vanes of 2ϵ, it can be readily shown that the residual attenuation is a minimum when the central vane is in an angular position that is mid-way between the angular positions of the two end vanes.) Let the voltage transmission coefficient of the central vane, for a wave component that is parallel to it, be denoted by $ae^{j\xi}$, where a is the modulus of the transmission coefficient and ξ is the differential phase change that occurs on transmission. Then, from the vector diagrams in Figs. A4.2a and A4.2b, it is seen that the output amplitude is given by

$$E_{out} = E_{in}\{\cos(\theta-\epsilon)\cos(\theta+\epsilon)+ae^{j\xi}\sin(\theta-\epsilon)\sin(\theta+\epsilon)\}$$
(A4.1)

Let

$$\cos(\theta-\epsilon)\cos(\theta+\epsilon) = M \qquad (A4.2)$$

and let

$$a\sin(\theta-\epsilon)\sin(\theta+\epsilon) = N \qquad (A4.3)$$

Then, the attenuation is

$$A'' = 20\log_{10}\frac{|E_{in}|}{|E_{in}(M+Ne^{j\xi})|} \qquad (A4.4)$$

$$= 20\log_{10}\frac{1}{\sqrt{\{M^2+N^2+2MN\cos\xi\}}} \qquad (A4.5)$$

Expanding eqn. A4.2, we get

$$M = (\cos\theta\cos\epsilon+\sin\theta\sin\epsilon)(\cos\theta\cos\epsilon-\sin\theta\sin\epsilon)(A4.6)$$

Combined effect of end vane misalignment   **299**

When ϵ is very small, $\cos \epsilon \approx 1$ and $\sin \epsilon \approx \epsilon$. Thus,

$$M \approx (\cos \theta + \epsilon \sin \theta)(\cos \theta - \epsilon \sin \theta)$$
$$= \cos^2 \theta - \epsilon^2 \sin^2 \theta \qquad (A4.7)$$

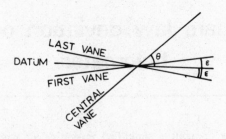

Fig. A4.1 Diagram defining the angles used in Appendix 4

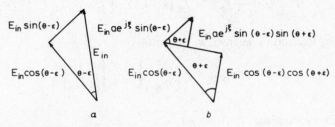

Fig. A4.2 Central vane vectors
    a Vectors at input end of central vane
    b Vectors at output end of central vane

Expanding eqn. A4.3, we get

$$N = a\{(\sin \theta \cos \epsilon - \cos \theta \sin \epsilon)(\sin \theta \cos \epsilon + \cos \theta \sin \epsilon)\}$$
$$\approx a\{(\sin \theta - \epsilon \cos \theta)(\sin \theta + \epsilon \cos \theta)\} \qquad (A4.8)$$
$$= a(\sin^2 \theta - \epsilon^2 \cos^2 \theta) \qquad (A4.9)$$

If the attenuation in dB produced by the central vane is denoted by A_v, we have

$$A_v = 20 \log_{10} \frac{1}{a}$$

Thus,

$$N = 10^{-(A_v/20)}(\sin^2 \theta - \epsilon^2 \cos^2 \theta) \qquad (A4.10)$$

The family of curves shown in Fig. 4.9 was obtained using eqns. A4.5, A4.7 and A4.10.

Square law deviation of a bolometer

This Appendix contains a simplified version of an analysis given by Sorger and Weinschel.[8] From eqn. 8.5, the total voltage across the bolometer is seen to be

$$V = I_B R = I_B \{R_0 + J(p + I_B^2 R)^n\} \tag{A5.1}$$

Using Maclaurin's theorem, we can write

$$V = V_{p=0} + p \left\{ \frac{\partial V}{\partial p_{p=0}} \right\} + \frac{p^2}{2} \left\{ \frac{\partial^2 V}{\partial p_{p=0}^2} \right\} \tag{A5.2}$$

Differentiating eqn. A5.1, we get

$$\frac{\partial V}{\partial p} = I_B \frac{\partial R}{\partial p} = I_B J n (p + I_B^2 R)^{n-1} \left(1 + I_B^2 \frac{\partial R}{\partial p}\right) \tag{A5.3}$$

Solving for $\partial R / \partial p$ we find that

$$\frac{\partial V}{\partial p} = \frac{I_B J n}{(p + I_B^2 R)^{1-n} - J n I_B^2} \tag{A5.4}$$

Differentiating again, we get

$$\frac{\partial^2 V}{\partial p^2} = -\frac{I_B J n (1-n)(p + I_B^2 R)^{-n} \left(1 + I_B^2 \dfrac{\partial R}{\partial p}\right)}{\{(p + I_B^2 R)^{1-n} - J n I_B^2\}^2} \tag{A5.5}$$

$$= -\frac{I_B J n (1-n)(p + I_B^2 R)^{1-2n}}{\{(p + I_B^2 R)^{1-n} - J n I_B^2\}^3} \tag{A5.6}$$

The first term in eqn. A5.2 is a d.c. component which is removed by the tuned amplifier. Thus, on comparing eqns. 8.1 and A5.2, it is seen that

$$\frac{U}{T} = \frac{\frac{1}{2}\left\{\frac{\partial^2 V}{\partial p_p^2 = 0}\right\}}{\left\{\frac{\partial V}{\partial p_p = 0}\right\}} \tag{A5.7}$$

From eqns. A5.4 and A5.6, we now get

$$\frac{U}{T} = -\frac{(1-n)(I_B^2 R)^{1-2n}}{2\left\{(I_B^2 R)^{1-n} - JnI_B^2\right\}^2} \tag{A5.8}$$

$$= -\frac{(1-n)}{2(I_B^2 R)}\left\{\frac{R}{R - Jn(I_B^2 R)^n}\right\}^2 \tag{A5.9}$$

Using eqns. A5.1 and A5.9, we finally get

$$\frac{U}{T} \approx -\frac{(1-n)}{2(I_B^2 R)}\left(\frac{R}{R_0}\right)^2 \tag{A5.10}$$

since $p \ll I_B^2 R$ and n is close to unity.

Error due to noise when measuring attenuation with a modulated sub-carrier system

When the attenuator under test is set at zero, let the amplitude modulated wave which reaches the mixer be denoted by

$$E(1 + m \cos \omega_m t) \cos \omega_c t$$

If the attenuator under test is now adjusted to give an attenuation of A dB, the wave arriving at the mixer is given by

$$E \cdot 10^{-A/20} (1 + m \cos \omega_m t) \cos \omega_c t$$

Let the ratio of the peak i.f. output voltage to the peak value of the modulation on the envelope be denoted by K. Then, the r.m.s. value of the i.f. voltage at the mixer output is given by

$$V_{if} = \frac{KEm \cdot 10^{-(A/20)}}{\sqrt{2}} \qquad (A6.1)$$

Differentiating and taking the modulus, we get

$$|\Delta V_{if}| = \frac{KEm}{8 \cdot 686 \sqrt{2}} 10^{-(A/20)} \Delta A \qquad (A6.2)$$

The total noise power referred to the mixer input is FkT_0B, where F is the overall noise factor, k is Boltzmann's constant, T_0 is 290K and B is the equivalent noise bandwidth of the receiver. Let us assume that the mixer matches the input waveguide which has a characteristic impedance of Z_0. Then, the r.m.s. noise voltage at the mixer output is

$$\Delta V_n = K \sqrt{FkT_0BZ_0} \qquad (A6.3)$$

The r.m.s. error in the attenuation measurement due to noise can be found by equating eqns. A6.2 and A6.3. Thus, replacing ΔA by ΔA_n, we get

$$\Delta A_n = \frac{8 \cdot 686 \sqrt{2FkT_0 BZ_0}}{Em} \, 10^{(A/20)} \tag{A6.4}$$

When the attenuator under test is set at zero, the total power that is fed into the mixer via the subcarrier channel is

$$P_0 = \frac{\omega_m}{2\pi} \int_{t=0}^{t=\frac{2\pi}{\omega_m}} \frac{E^2}{2Z_0} \, (1 + m \cos \omega_m t)^2 \, dt$$

$$= \frac{E^2}{2Z_0} \left(1 + \frac{m^2}{2}\right) \tag{A6.5}$$

The PSD yields the same signal to noise ratio that one would get if the i.f. equivalent noise bandwidth were made equal to the equivalent noise bandwidth of the PSD.[9] When the low pass filter in the PSD takes the form of a simple resistance capacitance filter of time constant τ,

$$B = \int_0^\infty \frac{df}{1 + 4\pi^2 f^2 \tau^2} = \frac{1}{4\tau} \tag{A6.6}$$

From eqns. A6.4, A6.5 and A6.6, we get

$$\Delta A_n = 4 \cdot 343 \sqrt{\frac{FkT_0}{P_0 \tau} \left\{\frac{1}{m^2} + \frac{1}{2}\right\}} \cdot 10^{(A/20)} \tag{A6.7}$$

Bilinear transformation of a circle in the complex plane

The bilinear transformation (also termed linear fractional transformation, homographic transformation and Möbius transformation) has the form $w = (az + b)/(cz + d)$ where all the symbols are, in general, complex numbers, and it represents in the w-plane a transformation of points in the z-plane. The object here is to determine the locus of points w which are related by the bilinear transformation to points z which lie on a circle of radius r with centre at $z = 0$. Thus, the z-locus is $z = re^{j\theta}$. No derivation of an expression of the w-locus *in a suitable form* for use in further analysis appears to be available, so the following derivation has been developed.

As usual, the transformation is best re-arranged as follows:

$$w = \frac{a\left(z + \dfrac{d}{c}\right) - \dfrac{ad}{c} + b}{c\left(z + \dfrac{d}{c}\right)} = \frac{a}{c} + \frac{(bc - ad)}{c^2} \cdot \frac{1}{\left(z + \dfrac{d}{c}\right)} \quad \text{(A7.1)}$$

which shows clearly that w is the resultant of the three following distinct well-known operations:

(1) reciprocal of $\left(z + \dfrac{d}{c}\right)$

(2) multiplication of reciprocal by the complex number $M = \left(\dfrac{bc - ad}{c^2}\right)$

(3) displacement or translation of the result (2) by the complex number $D = a/c$.

The simplest way to obtain expressions defining the locus of w is to note the effect of the operations (1), (2) and (3) above on a pair of extreme points of $(z + d/c)$ and then to apply eqns. A7.1 to the two extreme points. The three operations are as follows:

(1) The complex reciprocal of a circle with centre displaced from the origin, and not touching the origin, is known to be another circle of different radius displaced differently from the origin. The function $w = 1/z$ represents in the w-plane a *geometrical inversion* of points in the z-plane, relative to a circle of inversion of unit radius, together with a *reflection* in the real axis. This follows from the polar form $z = |z| e^{j\theta}$ which gives immediately $w = (1/|z|) e^{-j\theta}$ in which the $1/|z|$ is the inversion and the change from $e^{j\theta}$ to $e^{-j\theta}$ is the reflection in the real axis. The transformation of the circle BCD with centre at A, representing $(z + d/c)$ with $z = re^{j\theta}$, into the circle B'C'D' with centre at A', representing $1/(z + d/c)$, is shown in Fig. A7.1.

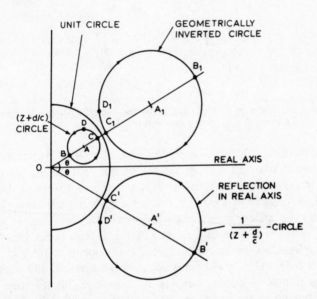

Fig. A7.1 Transformation of circle BCD representing $(z + d/c)$ into the circle B'C'D' representing $1/(z + d/c)$

As well as the case (*a*) shown where the circle BCD does not touch the origin 0, two other less common cases can occur:

(*b*) circles touching 0 are transformed into straight lines not touching 0, and vice versa

(*c*) straight lines through 0 are transformed into straight lines through 0

In Fig. A7.1, note that the minimum point B corresponds with the maximum point B' of the reciprocal or inverted circle, while the maximum point C corresponds with the minimum point C' of the reciprocal circle. Minimum and maximum refer to radial distances from 0.

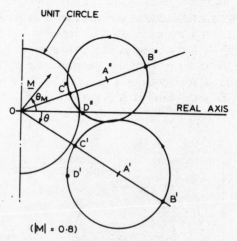

UNIT CIRCLE

REAL AXIS

$(|M| = 0.8)$

Fig. A7.2 Multiplication of circle B'C'D' by the complex number M shown by the vector $ML\theta_m$, giving the circle B"C"D"

(2) Multiplication by the complex number $M = |M|e^{j\theta}m$ consists of a 'magnification' of amount $|M|$ which does not distort the circle ($|M|$ magnifies distances from 0), together with a rotation about 0 of the magnified figure by the angle θ_M. The operation clearly does not affect the minimum and maximum properties of points C" and B", respectively, as radial distances from 0. See Fig. A7.2.

(3) Displacement by the complex number D amounts to a simple translation of the circle without rotation, by a distance $|D|$ in a direction θ_D relative to the real axis. Fig. A7.3 applies. This operation does not result in the transformed C" and B" remaining the minimum and maximum radial distances from 0 of points on the circle. Nevertheless, the points C''' and B''' are distinct and it is seen from the figure that the radius A'''B''' of the circle is half the modulus of the vectorial difference $(OB''' - OC''')$ and the position A''' of the centre is given by the vector (complex number) $OA''' = \frac{1}{2}(OB''' + OC''')$. It can now be seen that the radius and position of centre of the w-circle are simply derived from the maximum (C) and minimum (B) values of $z + d/c$ on the z-circle locus. Refer to Fig. A7.1 where $AC = r$ and $OA = d/c$, (a vector). Point C corresponds to $(d/c + r)$ and point B corresponds to $(d/c - r)$ where r

is real and both expressions relate to the direction OA of the vector d/c. Unit vector in this direction can be expressed $(d/c)/|(d/c)|$ which is used because it is in terms of the original complex constants. Then

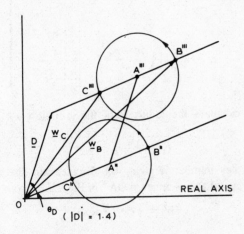

Fig. A7.3 Displacement of the circle B"C"D" by the complex number D shown by the vector $D\angle\theta_D$, giving the circle B'"C'"D'"

$$\left.\begin{array}{l} \left(\dfrac{d}{c}+z\right)_C = \dfrac{(d/c)}{|d/c|}\cdot\left[\left|\dfrac{d}{c}\right|+r\right] \\[3ex] \left(\dfrac{d}{c}+Z\right)_B = \dfrac{(d/c)}{|d/c|}\cdot\left[\left|\dfrac{d}{c}\right|-r\right] \end{array}\right\} \quad\text{(A7.2)}$$

and

$$\left.\begin{array}{l} \dfrac{1}{\left(\dfrac{d}{c}+z\right)_C} = \dfrac{|d/c|\,(c/d)}{\left(\left|\dfrac{d}{c}\right|+r\right)} \\[4ex] \dfrac{1}{\left(\dfrac{d}{c}+z\right)_B} = \dfrac{|d/c|\,(c/d)}{\left(\left|\dfrac{d}{c}\right|-r\right)} \end{array}\right\} \quad\text{(A7.3)}$$

Then, with eqn. A7.3 in eqn. A7.1 we have

$$
w_C = \frac{a}{c} + \left(\frac{bc - ad}{c^2}\right) \frac{\left|\dfrac{d}{c}\right| \dfrac{c}{d}}{\left|\dfrac{d}{c}\right| + r}
$$

$$
w_B = \frac{a}{c} + \left(\frac{bc - ad}{c^2}\right) \cdot \frac{\left|\dfrac{d}{c}\right| \dfrac{c}{d}}{\left|\dfrac{d}{c}\right| - r}
$$

$$(A7.4)$$

We have to calculate the radius R of the w-circle ($= A'''B'''$ in Fig. A7.3) from

$$R = \tfrac{1}{2} |w_B - w_C| \qquad (A7.5)$$

and the complex number W giving the distance and direction from 0 of the centre A''' of the w-circle ($= OA'''$ in Fig. A7.3) from

$$W = \tfrac{1}{2}(w_B + w_C) \qquad (A7.6)$$

Now

$$
\left[\frac{1}{\left|\dfrac{d}{c}\right| - r} + \frac{1}{\left|\dfrac{d}{c}\right| + r} \right] = \frac{2\left|\dfrac{d}{c}\right|}{\left|\dfrac{d}{c}\right|^2 - r^2}
$$

and

$$
\left[\frac{1}{\left|\dfrac{d}{c}\right| - r} - \frac{1}{\left|\dfrac{d}{c}\right| + r} \right] = \frac{2r}{\left|\dfrac{d}{c}\right|^2 - r^2}
$$

and with eqn. A7.4 in eqns. A7.5 and A7.6 we obtain

$$
R = \left| \left(\frac{bc - ad}{c^2}\right) \cdot \frac{\left|\dfrac{d}{c}\right| \dfrac{c}{d} \cdot r}{\left|\dfrac{d}{c}\right|^2 - r^2} \right| = r \left| \frac{bc - ad}{|d|^2 - r^2 |c|^2} \right| \qquad (A7.7)
$$

and

$$
W = \frac{a}{c} + \left(\frac{bc - ad}{c^2}\right) \frac{\dfrac{c}{d}\left|\dfrac{d}{c}\right|^2}{\left|\dfrac{d}{c}\right|^2 - r^2}
$$

$$= \frac{\dfrac{a}{c}\left|\dfrac{d}{c}\right|^2 - \dfrac{a}{c}r^2 + \dfrac{b}{d}\left|\dfrac{d}{c}\right|^2 - \dfrac{a}{c}\left|\dfrac{d}{c}\right|^2}{\left|\dfrac{d}{c}\right|^2 - r^2}$$

or

$$W = \frac{\dfrac{b}{d}\left|\dfrac{d}{c}\right|^2 - \dfrac{a}{c}r^2}{\left|\dfrac{d}{c}\right|^2 - r^2}.$$

Now $cc^* = |c|^2$ so $c^* = |c|^2/c$ and similarly $d^* = |d|^2/d$. Then

$$W = \frac{b\dfrac{|d|^2}{d} - a\dfrac{|c|^2}{c}r^2}{|d|^2 - r^2|c|^2} = \frac{bd^* - r^2ac^*}{|d|^2 - r^2|c|^2} \tag{A7.8}$$

Collecting results, the radius of the w-circle is

$$R = r\left|\frac{bc - ad}{|d|^2 - r^2|c|^2}\right|$$

and the position of its centre in the complex plane, relative to the origin, is

$$W = \frac{bd^* - r^2ac^*}{|d|^2 - r^2|c|^2}$$

In many applications, $r = 1$ and the circle in the z-plane is of unity radius.

Theory of weakly connected superconducting rings

A8.1 Introduction

Enough information is given in this Appendix to make the subject of the title plausible. Additional information is included to enable the interested reader to pursue the subject further and also to fill-in some gaps which, in the writer's opinion, exist in the available literature. To understand and describe adequately the phenomena of super-conductivity and superconductive tunnelling, a knowledge of quantum mechanics is required. This is the mechanics of processes on an atomic scale and because things on such a small scale behave like nothing found in everyday 'classical' physics, its rules can only be inferred from observation of how nature behaves on an ultimately small scale. Thus, the theory of quantum mechanics could only be evolved from a series of brilliant ideas based on observations of 'how it is' in nature. A brief and authoritative account of how the theory as at present accepted evolved (up to 1930) is still available in the appendix on 'Mathematical apparatus' in Heisenberg's book[10] 'The physical principles of the quantum theory'. The axiomatic presentation of the foundations, which is found in many good books published in the last 40 years, makes for a tidy presentation and is a good way to learn the subject, but it should be borne in mind that this approach can give the wrong impression that the theory was or could be developed by purely intellectual arguments on a classical foundation, with only slight modification.

As a result of people trying to describe quantum phenomena in terms of classical concepts, widespread confusion in interpretation of the theory has arisen. At the root of most of the trouble has been confusion of the complex probability amplitude in configuration space with a classical physical wave amplitude in 3-dimensional real space. Schrödinger's equation does not deal with matter waves but with probability amplitudes. It is important to remember this in the theory of superconductivity because that subject deals with quantum phenomena which are observable on a macroscopic scale and people have been tempted to regard Schrödinger waves there as classical waves.

A good modern introductory text which takes a sensible view is found in Feynman's Lectures on Physics,[11] Vol. 3. This text concludes with a chapter on superconductivity which is outstanding in its simplicity and is much quoted.

A8.2 Generalised momentum in a magnetic field

Before proceeding with the theory of the superconducting ring it is necessary to establish certain classical relations between energy, kinetic momentum and total momentum which depend on the magnetic vector potential. These relations have received scant treatment in most texts. The force on a charged particle in motion is the sum of the forces of the electric field intensity E and the magnetic flux density B. The first force is eE and the second is $e(v \times B)$ where v is the particle velocity and e is its charge (positive for a positive charge). Thus, the equation of motion for the charged particle of mass m is the Lorentz equation

$$\frac{d}{dt}(mv) = e(E + v \times B) \tag{A8.1}$$

Now from electromagnetic theory

$$B = \operatorname{curl} A \tag{A8.2}$$

where A is the magnetic vector potential, so the Maxwell equation $\operatorname{curl} E = -\partial B/\partial t$ becomes $\operatorname{curl} E + \partial/\partial t \,(\operatorname{curl} A) = 0$ or because curl is a linear operator, $\operatorname{curl}(E + \partial A/\partial t) = 0$. It follows that the vector $E + \partial A/\partial t$ must be the gradient of some scalar, say $-\operatorname{grad} \phi$. So

$$E = -\operatorname{grad} \phi - \partial A/\partial t \tag{A8.3}$$

Putting eqns. A8.2 and A8.3 into eqn. A8.1:

$$\frac{d}{dt}(mv) = e\left\{-\operatorname{grad} \phi - \frac{\partial A}{\partial t} + (v \times \operatorname{curl} A)\right\} \tag{A8.4}$$

We require to express $(v \times \operatorname{curl} A)$ more conveniently and for simplicity we first take the x-component, which is $\dot{y}\,(\operatorname{curl} A)_z - \dot{z}\,(\operatorname{curl} A)_y$ where $\dot{x}$ means dx/dt and similarly for y and z. Then in detail

$$(v \times \text{curl } A)_x = \dot{y}\left(\frac{\partial A_y}{\partial x} - \frac{\partial A_x}{\partial y}\right) - \dot{z}\left(\frac{\partial A_x}{\partial z} - \frac{\partial A_z}{\partial x}\right).$$

Add and subtract $\dot{x}\partial A_x/\partial x$. Then we have for the x-component:

$$\dot{x}\frac{\partial A_x}{\partial x} + \dot{y}\frac{\partial A_y}{\partial x} + \dot{z}\frac{\partial A_z}{\partial x} - \left(\dot{x}\frac{\partial A_x}{\partial x} + \dot{y}\frac{\partial A_x}{\partial y} + \dot{z}\frac{\partial A_x}{\partial z}\right)$$

$$= \frac{\partial}{\partial x}(\dot{x}A_x + \dot{y}A_y + \dot{z}A_z) - \left(\dot{x}\frac{\partial}{\partial x} + \dot{y}\frac{\partial}{\partial y} + \dot{z}\frac{\partial}{\partial z}\right)A_x$$

because $\partial \dot{x}/\partial x = 0$ etc. The y- and z-components are obvious from symmetry considerations and the expressions in brackets are clearly scalar products. So, adding the 3 components with their respective unit vectors a, b and c included as factors, we have

$$v \times \text{curl } A = \text{grad } (v \cdot A) - (v \cdot \nabla)A \tag{A8.5}$$

where

$$v = a\dot{x} + b\dot{y} + c\dot{z} \text{ and } \nabla = a\frac{\partial}{\partial x} + b\frac{\partial}{\partial y} + c\frac{\partial}{\partial z}$$

We also note that

$$\frac{dA}{dt} = \frac{\partial A}{\partial t} + (v \cdot \nabla)A \tag{A8.6}$$

with components

$$\frac{dA_x}{dt} = \frac{\partial A_x}{\partial t} + \left(\dot{x}\frac{\partial}{\partial x} + \dot{y}\frac{\partial}{\partial y} + \dot{z}\frac{\partial}{\partial z}\right)A_x \text{ etc.,}$$

is the rate of change of A experienced when moving with the particle. $\partial A/\partial t$ is the *local* rate of change and $(v \cdot \nabla)A$ is the convective rate of change.

With eqns. A8.5 and A8.6 in eqn. A8.4 and some re-arrangement

$$\frac{d}{dt}(mv + eA) = e \text{ grad } (v \cdot A - \phi) \tag{A8.7}$$

which is an alternative form of eqn. A8.1 with the *total* momentum

$$p = mv + eA \tag{A8.8}$$

and the total force expressed as the gradient of a scalar. Here p is the total momentum, mv is the *kinetic* momentum and eA is the *electromagnetic* momentum.

The *energy* of the particle in the field is given by the line integral of the force over the path of the particle. It is simpler to use eqn. A8.1 with eqn. A8.3 for E than to use eqn. A8.7. Both, of course, give the same result. So if ds is an infinitesimal vector in the direction of the path at each point, we have

$$\int \frac{d}{dt}(mv) \cdot ds + e \int \mathrm{grad}\; \phi \cdot ds + e \int \frac{\partial A}{\partial t} \cdot ds - e \int (v \times B) \cdot ds$$

$$= \text{constant} = W(\text{say})$$

For the 1st term, take the x-component first and find:

$$m \int \frac{d\dot{x}}{dt} \cdot dx = m \int \frac{d\dot{x}}{dt} \dot{x}\, dt = m \int \dot{x}\, d(\dot{x}) = \tfrac{1}{2} m \dot{x}^2$$

So

$$\int \frac{d}{dt}(mv) \cdot ds = \tfrac{1}{2} m (\dot{x}^2 + \dot{y}^2 + \dot{z}^2) = T$$

the *kinetic energy*. For the 2nd term we find $e\phi$ and the 3rd term cannot be reduced much. In the 4th term, $(v \times B)$ is perpendicular in direction to the plane containing v and B and ds is in the direction of v. So the scalar product of $(v \times B)$ and ds is zero, hence the 4th integral is taken as zero. In physical terms, the magnetic force does no work on the moving charged particle because the motion is always at right angles to the magnetic force. We are left with

$$W = T + e\phi + e\frac{\partial}{\partial t} \int A \cdot ds \qquad (A8.9)$$

as the energy of the particle. If the magnetic field is constant in time, $W = T + e\phi$ and the magnetic field does no more than curve the path which otherwise would be determined by ϕ as a function of position. The last term in eqn. A8.9 is that part of the potential energy associated with the familiar e.m.f. arising from the rate of change of magnetic flux. [In fact, if the integration is around a complete contour.

$$\oint A \cdot ds = \iint B \cdot dS = \Phi$$

where dS is an element of surface and Φ is the total flux threading the contour.]

The *Hamiltonian* function is of importance in formulating a mechanical problem classically so that it can be reformulated in quantum theory, where such an analogy is possible. First, however, a brief acquaintance with the Lagrangian function and equations is necessary. For simplicity, we confine ourselves to the Cartesian coordinates of one particle. The x-component of the equation of motion is $d/dt\,(m\dot{x}) = F_x$, and, in terms of the kinetic energy, $T = \tfrac{1}{2} m (\dot{x}^2 + \dot{y}^2 + \dot{z}^2)$: one may write $\partial T/\partial \dot{x} = m\dot{x}$ for the x-component of momentum.

So with the x-component of force F_x:

$$\frac{d}{dt}\left(\frac{\partial T}{\partial \dot{x}}\right) = F_x \tag{A8.10}$$

If F_x is the gradient of a scalar potential, say $F_x = -\partial V/\partial x$ then

$$\frac{d}{dt}\left(\frac{\partial T}{\partial \dot{x}}\right) + \frac{\partial V}{\partial x} = 0$$

or if we define $\mathcal{L} = T - V$ (the Lagrangian) where $\mathcal{L}$ is a function of $x, y, z; \dot{x}, \dot{y}, \dot{z}$ explicitly, then

$$\frac{d}{dt}\left(\frac{\partial \mathcal{L}}{\partial \dot{x}}\right) - \frac{\partial \mathcal{L}}{\partial x} = 0 \tag{A8.11}$$

This is the standard Lagrange form of a component equation of motion. In the more general case, which we need, where F_x may depend on $\dot{x}$ as well as x, we define a new function $\mathcal{L}_f$ so that

$$\frac{d}{dt}\left(\frac{\partial \mathcal{L}_f}{\partial \dot{x}}\right) - \frac{\partial \mathcal{L}_f}{\partial x} = -F_x \tag{A8.12}$$

Then we define $\mathcal{L} = T + \mathcal{L}_f$. From eqn. A8.12 with eqn. A8.10 we obtain an expression like eqn. A8.11 again. If we now substitute the right-hand side of eqn. A8.4 for F_x and then use eqns. A8.5 and A8.6 we obtain for the x-component

$$\frac{d}{dt}\left(\frac{\partial \mathcal{L}_f}{\partial \dot{x}}\right) - \frac{\partial \mathcal{L}_f}{\partial x} = e\left\{\frac{dA_x}{dt} + \frac{\partial}{\partial x}\left[\phi - (\mathbf{v} \cdot \mathbf{A})\right]\right\}$$

and

$$\mathcal{L}_f = -e\left[\phi - (\mathbf{v} \cdot \mathbf{A})\right]$$

will be found to satisfy this. So the *Lagrangian* function for a charged particle in electric and magnetic fields is expressed

$$\mathcal{L} = T - e\left[\phi - (\mathbf{v} \cdot \mathbf{A})\right] \tag{A8.13}$$

To proceed to the *Hamiltonian*, we rewrite eqn. A8.11 in generalised coordinates $q_r (r = 1$ to 3 for a single particle without constraint)

$$\frac{d}{dt}\left(\frac{\partial \mathcal{L}}{\partial \dot{q}_r}\right) - \frac{\partial \mathcal{L}}{\partial q_r} = 0 \tag{A8.14}$$

where $\mathcal{L} = \mathcal{L}(q, \dot{q}, t)$ in general. By analogy with the Newtonian equations of motion we define

$$p_r = \frac{\partial \mathcal{L}}{\partial \dot{q}_r} \tag{A8.15}$$

as generalised components of momenta. Then with eqn. A8.13 put in eqn. A8.15 we obtain (with Cartesian coordinates)

$$p_x = m\dot{x} + eA_x \text{ etc.} \tag{A8.16}$$

which agrees with eqn. A8.8. It can be shown[12] that the quantity $\mathcal{H}$ which depends explicitly on q, p and t in general, defined by

$$\mathcal{H}(q, p, t) = \sum_r p_r \dot{q}_r - \mathcal{L}(q, \dot{q}, t) \tag{A8.17}$$

is constant if $\partial\mathcal{L}/\partial t = 0$, i.e. if the Lagrangian does not depend explicitly on t. In fact, $d\mathcal{H}/dt = \partial\mathcal{H}/\partial t = -\partial\mathcal{L}/\partial t$ and the equations of motion (termed Hamilton's canonical equations) are 1st-order differential equations:

$$\dot{q}_r = \partial\mathcal{H}/\partial p_r \quad \text{and} \quad \dot{p}_r = -\partial\mathcal{H}/\partial q_r \tag{A8.18}$$

From eqn. A8.17 with eqn. A8.15 our example of a particle in an electromagnetic field gives:

$$\mathcal{H} = \sum (m\dot{x} + eA_x\dot{x}) - \frac{1}{2}\sum (m\dot{x}^2) + e\phi - e\sum (xA_x)$$

$$= \frac{1}{2m}\sum (m\dot{x})^2 + e\phi$$

So from eqn. A8.16 the Hamiltonian for a charged particle moving in an electromagnetic field is

$$\mathcal{H} = \frac{1}{2m}\sum (p_x - eA_x)^2 + e\phi \tag{A8.19}$$

It can clearly be formed from the Hamiltonian in a field with no magnetic force by substituting $(p_x - eA_x)$ for p_x, or more correctly, by noting that $\mathcal{H} = (1/2m)(mv)^2 + e\phi$ always and that $mv = p - eA$ vectorially.

The importance of the Hamiltonian to us is that in quantum mechanics Schrödinger's equation can be obtained from it simply by substituting the operator $-j\hbar\nabla$ for p (generalised momentum), $j\hbar\partial/\partial t$ for W (energy) and operating by all terms of $\mathcal{H}$ on the probability amplitude function ψ (or in the more explicit Dirac notation, $\langle q \mid W \rangle$).

When the magnetic field, through A is constant in time, W and $\mathcal{H}$ are identical. But when $\partial A/\partial t \neq 0$, we have from eqns. A8.9 and A8.19

$$W = T + e\phi + e\frac{\partial\Phi'}{\partial t} \quad \text{and} \quad \mathcal{H} = T + e\phi$$

where $\Phi' = \int A \cdot ds$, and the difference between W and $\mathcal{H}$ is the potential energy due to the explicit rate of change of magnetic flux in the path of the particle.

This explicit time variation of magnetic flux occurs in weakly-connected superconducting rings with high-frequency alternating magnetic fields applied, the very problem in which we are interested.

A8.3 Superconductivity

An elementary and up to date account of superconductivity can be found in the book by Rose-Innes and Rhoderick.[13] The first point noted here is that superconductivity does not just involve zero resistance to steady currents at temperatures below a critical value (for most of the superconducting metals, below 10K), but a metal in a superconducting state expels all magnetic flux from its interior except in a thin skin at the surface where the supercurrents are confined. (The Meissner effect.) In fact, this effect is explained by the existence of a 'screening' supercurrent at the surface of the metal, which produces just enough magnetic flux to cancel that inside the metal produced by an external source.

Secondly, there is a limiting current density above which the superconducting state reverts to the normal state. If the applied magnetic field is increased to the point where the screening current results in a current density which exceeds the critical value, then the state again reverts to normal. This is usually interpreted as a 'critical magnetic field strength', characteristic of the particular metal, as is the critical current density.

These things are not explainable classically. They are, in fact, quantum phenomena which are unusual in that they are directly observable on a classical scale. An elementary outline of the quantum theory of the superconductive state is given in Chapter 9 of Reference 13. A more mathematical account is given in Reference 19.

As a result of quantum interaction between the conduction electrons and vibrations of the lattice formed by the nuclei and the bound electrons, there is a weak but effective attractive force between conduction electrons with opposite spins, weakly binding them into pairs. The force is effective only at very low temperatures where the thermal agitation is too small to break up the electron-pairs. This pairing occurs on an average in the statistical sense, and most probably the two electrons have opposite and equal momenta relative to the centre of mass of the pair. Motion of the centre of mass is regarded as motion of the pair, often termed a 'Cooper-pair'. One is warned not to take this particle picture of a 'bound pair' too literally in the classical sense. The whole thing is a quantum phenomenon (see Section A8.1).

In the *normal* state of a metal, the conduction electrons are 'Fermi' particles[11] of which only one can be in a given quantum state at a time. (In a system of many particles with weak interaction, there are more than enough states available to accommodate the many electrons.) Under this condition, moving electrons frequently become scattered hence the resistance which arises from the limitation of the mean velocity of these electrons.

In the *superconductive* state of a metal, the paired conduction electrons are 'Bose' particles[11] which all tend strongly to be in the same state, that is, they all tend to move with the same momentum. The only scattering process which can limit or reduce the current is when some of the pairs are broken up into separate electrons. It is helpful to think of the Schrödinger probability amplitude wave for the Cooper-pairs, because as all pairs tend to be in one state, the system of the pairs in a given superconductor can be described by a single elementary Schrödinger equation, without spin. This development is well presented by Feynman,[11] Chapter 21, where from such an equation he derives formulae for the d.c. skin effect, deduces the Meissner effect and formulates the quantisation of magnetic flux in superconducting rings.

A8.4 Superconducting rings

If a superconducting ring had been in a magnetic field before being lowered in temperature to below the critical value, the total magnetic flux threading the hole in the ring will remain constant at its original value so long as the ring remains in a superconducting state. Then, when the external source of magnetic flux is removed, the total flux threading the hole will not change, but the flux outside the ring will be simply that formed by the closed curves of the flux threading the hole. The flux-density distribution in the hole may change when the external source is removed, but the total flux in the hole is unchanged. Of course, all flux is expelled from inside the metal of the ring while it is in a superconducting state, and there will be boundary supercurrents at the surface of the ring to ensure this (Fig. A8.1).

The unchanging magnetic flux Φ which threads the ring turns out to be quantised in multiples of $|\Phi_0| = h/2|e| = 2 \cdot 068 \times 10^{-15}$ Wb. This flux quantum is just large enough to be observable experimentally, and this is of central importance in the application of superconductivity to measurement standards. So we shall show how it arises from the theory.

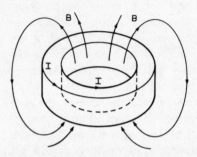

Fig. A8.1 Superconducting ring with magnetic field threading through hole in centre
Screening super-current *I* at the surface cancels any magnetic field inside the metal of the ring and ensures the persistence of the magnetic field shown

By formulating the Schrödinger equation from our eqn. A8.19 and writing the complex probability amplitude in the form $\psi = ue^{j\theta}$ with real magnitude u and phase θ, it is shown in Reference 11, Chapter 21, that the current density J in a superconductor with electron-pairs of charge $2e$ (where $e = -1\cdot602 \times 10^{-19}$ C) and mass $2m$, can be expressed

$$J = \frac{\hbar}{2m}\left(\operatorname{grad}\theta - \frac{2e}{\hbar}A\right)\rho \qquad (A8.20)$$

where the vector potential A defines the applied magnetic field and ρ is the volume density of charge in the superconductor due to the electron-pairs which form the current by their motion. $\hbar$ is Planck's constant $\div 2\pi$. From eqn. A8.20

$$\operatorname{grad}\theta = \frac{2m}{\hbar\rho}J + \frac{2e}{\hbar}A$$

and

$$\theta = \int \operatorname{grad}\theta \cdot ds = \frac{2m}{\hbar\rho}\int J \cdot ds + \frac{2e}{\hbar}\int A \cdot ds \qquad (A8.21)$$

where integration is along a path inside the metal of the superconducting ring. Eqn. A8.21 can also be derived by using eqn. A8.8 directly. The phase of the complex probability amplitude $\langle s | p \rangle$, where p is the generalised momentum and s is distance along the path of the particle, is $\theta = p \cdot s/\hbar$ or if p varies slowly with position s,

$$\theta = \frac{1}{\hbar} \int p \cdot ds \qquad (A8.22)$$

For an electron-pair $p = 2mv + 2eA$ from eqn. A8.8 and we have $J = \rho v$ so $v = J/\rho$. Then eqn. A8.21 follows from eqn. A8.22 immediately.

In order that the phase θ of the Schrödinger wave around the ring is *single valued*, the change in θ around the ring must be 0 or $\pm 2\pi n$ where n is an integer, so that there is a persistent wave. So from eqn. A8.21, multiplying throughout by $\hbar/2e$,

$$\pm 2\pi n \cdot \frac{\hbar}{2e} = \frac{m}{\rho e} \oint J \cdot ds + \oint A \cdot ds$$

Now by Stokes' theorem,

$$\oint A \cdot ds = \iint B \cdot ds$$

$$= \Phi \text{ (the total flux threading the contour)}$$

Also, we write $\Phi_0 = h/2e$. So:

$$n\Phi_0 = \frac{m}{\rho e} \oint J \cdot ds + \Phi \qquad (A8.23)$$

Now the current density J decays exponentially with distance into the metal, so if the contour is inside the metal ring where J is negligible, then Φ includes almost all the magnetic flux of the current in the metal, and that in the hole formed by the ring. So $n\Phi_0 = \Phi$ and the flux Φ is quantised in multiples of Φ_0. (The right hand side of eqn. A8.23 has been termed 'fluxoid' to distinguish it from 'flux' in the general case where J is appreciable on the contour chosen.)

What happens when the temperature is lowered to produce superconductivity when an applied flux Φ threads the ring, where $\Phi \neq n\Phi_0$? The answer is that the flux threading the ring jumps (up or down) to the nearest multiple of Φ_0.

The only way Φ can be changed is either for the ring to lose its superconductivity for a time, when the change can take place, or for the ring to be open-circuited in some controlled way. This is achieved by including a 'weak-link' in the ring, which has a critical current corresponding to flux of the same order as the flux quantum. Then the superconducting circuit is interrupted whenever the current through the weak link exceeds the critical value.

The most commonly used form of weak link is the Josephson junction, which is described theoretically in Reference 11, Chapter 21, and in References 14, 15 and 19. The behaviour of the Josephson junction as a circuit element, particularly at currents exceeding the critical value, is described in detail in Reference 15, Chapter 11.

A8.5 Weakly-connected superconducting rings

We are concerned only with a superconducting loop or ring containing one weak link (a Josephson junction), operated at radio frequency, in which the r.f. skin effect confines current to the surface of the metal irrespective of the superconducting d.c. skin effect. The form of the ring may be completely different from that shown in Fig. A8.1. The 'hole' in the centre of the 'ring' may take the form of an annular ring around a flat-plate gap to form a cavity (L-band), or take the form of a $\lambda_g/4$ length of low-impedance, short-circuited waveguide (X-band). In all cases, it has to be made of 'superconductive' metal, and be closed as a circuit by a Josephson junction, usually a niobium screw with pointed end pressing on a niobium anvil.

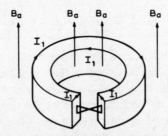

Fig. A8.2 Split-superconducting ring with weak-link bridging the gap
    Screening super-current I_1 keeps the field out of the metal of the ring.
    I_1 is distinct from any (usually much smaller) supercurrent I that flows
    through the weak link and the surface of the ring

Purely for illustration of principles and to aid understanding, an idealised form of ring with gap closed by a weak link is shown in Fig. A8.2. Externally applied magnetic flux of density B_a is shown. We suppose the gap is very narrow (unlike the Figure, which is to show it clearly) so the length of the weak link is small. If the weak link was removed, the screening supercurrent I_1 would flow over the surface as indicated, cancelling all flux inside the metal. It can be seen that I_1 does not have to flow through the weak link when it is replaced. Now, when the current through the weak link is below the critical value I_c and supercurrent can flow all round the hole in the ring (within the limits $\pm I_c$), the boundary conditions are such as to constrain the total fluxoid linking the ring to $n\Phi_0$ where n is zero or an integer. We then have an equation like eqn. A8.23 which here we write

$$n\Phi_0 = \frac{1}{2e} \int_1^2 p \cdot ds + \Phi \qquad (A8.24)$$

where $\Phi = LI$, where L is the inductance of the ring (less the small gap) and the integral is across the small gap 1, 2.

Let us assume that the weak link is an ideal Josephson junction at zero K, so there is no breaking up of Cooper pairs by thermal agitation and so no resistance shunting the junction. For such a junction, it can be shown[11,15] (especially Feynman,[11] Chapter 21) that the following holds:

$$I = I_c \sin \theta_{12} \qquad (A8.25)$$

where θ_{12} is the phase difference across the Josephson junction of the Schrödinger wave for the tunnelling electron pairs. I is the supercurrent passing and I_c is the critical current for the junction. Now

$$\theta_{12} = \frac{1}{\hbar} \int_1^2 p \cdot ds$$

as in eqn. A8.22, so with eqn. A8.25

$$\hbar \sin^{-1} \left(\frac{I}{I_c} \right) = \int_1^2 p \cdot ds$$

Then in eqn. A8.24

$$n\Phi_0 - \Phi = \frac{\hbar}{2|e|} \sin^{-1} \left(\frac{I}{I_c} \right) = \frac{\Phi_0}{2\pi} \sin^{-1} \left(\frac{I}{I_c} \right)$$

and

$$I = I_c \sin \left\{ \frac{2\pi}{\Phi_0} (n\Phi_0 - \Phi) \right\}$$

which reduces to

$$I = -I_c \sin \left(\frac{2\pi\Phi}{\Phi_0} \right) \qquad (A8.26)$$

The negative sign arises from the choice of positive Φ_0 direction. Now the total flux Φ threading the ring is clearly given by

$$\Phi = \Phi_a + LI \qquad (A8.27)$$

where Φ_a is the total *applied* magnetic flux threading the hole in the ring, and I is the supercurrent circulating around the hole in the ring and through the Josephson junction. L is the effective inductance of the ring to relate current to flux. With eqns. A8.26 and A8.27 we obtain the relations

$$\Phi = \Phi_a - LI_c \sin\left(\frac{2\pi\Phi}{\Phi_0}\right) \tag{A8.28}$$

$$LI = -LI_c \sin\left\{\frac{2\pi}{\Phi_0}(\Phi_a + LI)\right\} \tag{A8.29}$$

These are not explicit expressions for Φ and LI as functions of Φ_a, so it is better to express Φ_a as a function of Φ and plot curves for given LI_c values. The curves for LI are obtained from the Φ curves, by noting the difference between the line $\Phi = \Phi_a$ and the curve Φ.

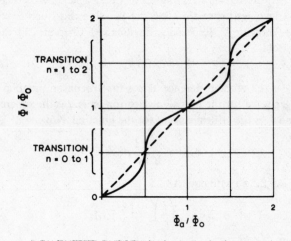

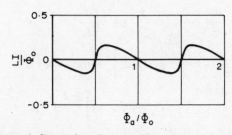

Fig. A8.3A Total magnetic flux Φ threading a weakly-connected ring as a function of applied magnetic flux Φ_a threading the ring, in units of flux quanta Φ_0
For the weak-link critical current $I_c = 0.159\,\Phi_0/L$ (where L is inductance of ring)

Fig. A8.3A shows the curves with $LI_c = \Phi_0/2\pi$ while Fig. A8.3B shows the curves with $LI_c = \Phi_0/\pi$. In the latter case, there is hysteresis where the curves are double-valued, which means that the transitions occur at slightly different values of Φ_a according to whether the applied flux Φ_a is increasing or decreasing.

The purpose of drawing these curves is to show what happens to the total flux Φ threading a weakly-connected superconducting ring when the applied flux Φ_a increases from zero to some value, and then may perhaps decrease again.

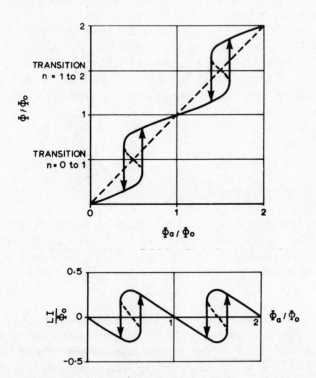

Fig. A8.3B Total magnetic flux Φ threading a weakly-connected ring as a function of applied magnetic flux Φ_a threading the ring, in units of flux quanta Φ_0
For the weak-link critical current $I_c = 0.318 \, \Phi_0/L$. Note hysteresis at transitions between quantum states

The fluxoid (eqn. A8.24) is strictly quantised so long as there is an unbroken ring of superconductor, and the foregoing shows that the flux Φ (slightly different from the fluxoid) follows a curve which approximates to the stepped function of the fluxoid. Differences between the line $\Phi = \Phi_a$ and the Φ-curve are made up by the variation in the current through the ring and Josephson junction, which varies between $\pm I_c$. When I_c is large enough to produce hysteresis, there is some slight loss experienced whenever Φ_a moves up and down in sufficient amount to span several quanta Φ_a.

The original paper on weakly-connected superconducting rings by Silver and Zimmerman[16] also treated the case of a simple weak link which does not show the phase-change of a Josephson junction but merely has a critical current value I_c. The resulting curves of Φ versus Φ_a are discontinuous and difficult to treat, but show all the same effects as the curves shown here for the Josephson junction used as a weak link. One point brought out in Reference 16 is that the time constant of a typical ring and weak link is of the order of 10^{-12} s, so that, for frequencies up to at least 10^{10} Hz, the steady-state characteristics can be applied to high-frequency problems without error, provided the e.m.f. arising from $\partial\Phi/\partial t$ is accounted for. This also means that measurements of flux (or of current producing the flux) using the device are likely to be quite independent of frequency within wide limits.

A8.6 Dynamic behaviour

Let the applied magnetic flux have a component that varies sinusoidally with time together with a component that is steady, or sinusoidal in time with a much lower frequency than the first component, so that it can be treated approximately as quasi-steady. Thus we write

$$\Phi_a = \Phi_1 + \Phi_2 \sin \omega t \qquad (A8.30)$$

This relation is precise when the weakly-connected ring is used as a magnetometer in which Φ_1 is the steady flux measured and Φ_2 is the amplitude of the r.f. applied flux (at about 30 MHz). The relation in the form of eqn. A8.30 can also be applied approximately to the problem of r.f. attenuation measurement in which Φ_1 has a v.h.f. component of angular frequency ω_s which is very much lower than the angular frequency ω. Typically, $f_s = 30$ MHz and $f = 9$ GHz. It has already been stated that the dynamic behaviour of the weakly-connected ring can be taken as similar to that of a slowly changing or quasi-static state except that the rate of change of Φ leads to an e.m.f. around the ring and *in any circuit inductively coupled to the ring*. Thus in the case of attenuation measurement, this e.m.f., which will have little pulses imposed on it at intervals of Φ_0 (see Fig. A8.3), will in effect impose similar 'markers' at intervals of Φ_0 on the voltage reflection coefficient of the termination of the 9 GHz waveguide provided by the ring. (See Fig. 12.12). We have for this e.m.f. in the ring

$$-V(t) = \frac{d\Phi}{dt} = \frac{d\Phi}{d\Phi_a} \cdot \frac{d\Phi_a}{dt}$$

and with eqn. A8.30

$$-V(t) = \omega \, \Phi_2 \, \frac{d\Phi}{d\Phi_a} \cos \omega t \qquad (A8.31)$$

Now eqn. A8.28 gives

$$\Phi_a = \Phi + LI_c \sin (2\pi\Phi/\Phi_0)$$

So

$$d\Phi_a/d\Phi = 1 + (2\pi LI_c/\Phi_0) \cos (2\pi\Phi/\Phi_0)$$

Since Φ differs from Φ_a by only a small amount, and since we can assert that LI_c/Φ_0 is small so that the second term in the previous expression is small compared with 1, then

$$\frac{d\Phi}{d\Phi_a} = \left(\frac{d\Phi_a}{d\Phi} \right)^{-1} \simeq 1 - \frac{2\pi LI_c}{\Phi_0} \cos \left(\frac{2\pi\Phi_a}{\Phi_0} \right)$$

or, if we substitute eqn. A8.30 for Φ_a, then

$$\frac{d\Phi}{d\Phi_a} \simeq 1 - A \cos a_1 \cdot \cos (a_2 \sin \theta) + A \sin a_1 \cdot \sin (a_2 \sin \theta)$$
$$(A8.32)$$

where

$$A = 2\pi LI_c/\Phi_0; \qquad a_1 = 2\pi\Phi_1/\Phi_0;$$

$$a_2 = 2\pi\Phi_2/\Phi_0 \text{ and } \quad \theta = \omega t;$$

and we have used the expression

$$\cos(a_1 + a_2 \sin \theta) = \cos a_1 \cos(a_2 \sin \theta) - \sin a_1 \sin(a_2 \sin \theta)$$

With (A8.32) substituted in (A8.31), we obtain

$$-V(t) = \omega\Phi_2 \{1 - A \cos a_1 \cdot \cos (a_2 \sin \theta) + A \sin a_1 \sin (a_2 \sin \theta)\} \cos \theta$$
$$(A8.33)$$

To find the amplitude of the fundamental for the frequency $\omega/2\pi$, we calculate the appropriate Fourier coefficients in

$$V(t) = V_0 + V_1' \cos \omega t + \ldots + V_1'' \sin \omega t + \ldots$$

which are

$$V_1' = \frac{1}{\pi} \int_0^{2\pi} V(t) \cos \theta \cdot d\theta \qquad (A8.34)$$

$$V_1'' = \frac{1}{\pi} \int_0^{2\pi} V(t) \sin \theta \cdot d\theta \qquad (A8.35)$$

We consider only the fundamental because in most microwave networks the response to harmonics is very low. It is tedious but not difficult to show that

$$\cos (a_2 \sin \theta) = J_0(a_2) + 2J_2(a_2) \cos 2\theta + 2J_4(a_2) \cos 4\theta + \ldots$$

$$\sin{(a_2 \sin\theta)} = 2J_1(a_2)\sin\theta + 2J_3(a_2)\sin 3\theta + \ldots$$

and using these equations in eqn. A8.33

$$V_1' = -\frac{\omega\Phi_2}{\pi} \int_0^{2\pi} \{1 - A\cos a_1 \, [J_0(a_2) + 2J_2(a_2)\cos 2\theta + \ldots]$$

$$+ A\sin a_1 \, [2J_1(a_2)\sin\theta + \ldots]\}(\tfrac{1}{2} + \tfrac{1}{2}\cos 2\theta)\cdot d\theta$$

where the factor $(\tfrac{1}{2} + \tfrac{1}{2}\cos 2\theta)$ replaces $\cos^2\theta$. Any term in the integrand depending on θ in the form $\sin n\theta$ or $\cos n\theta$ vanishes in the definite integral. We are left with:

$$V_1' = -\omega\Phi_2 \, \{1 - A\cos a_1 J_0(a_2) - A\cos a_1 J_2(a_2)\}$$

in which the relation $\cos^2 2\theta = \tfrac{1}{2} + \tfrac{1}{2}\cos 4\theta$ was used. The Bessel function recurrence formula

$$J_{n-1}(x) + J_{n+1}(x) = \frac{2n}{x}J_n(x)$$

gives

$$J_0(x) + J_2(x) = \frac{2}{x}J_1(x)$$

so

$$V_1' = -\omega\Phi_2 \, \{1 - (2A/a_2)\cos a_1 J_1(a_2)\}$$

and with the values of A, a_1 and a_2 inserted we finally obtain

$$V_1' = -\omega\Phi_2 + 2\omega LI_c \cos\left(\frac{2\pi\Phi_1}{\Phi_0}\right) J_1\left(\frac{2\pi\Phi_2}{\Phi_0}\right) \quad \text{(A8.36)}$$

Similar calculation shows that $V_1'' = 0$.

Eqn. A8.36 shows that the induced voltage, or the voltage reflection coefficient of a waveguide coupled to the ring, for the frequency $\omega/2\pi$, has a dependence on the quasi-steady or more slowly oscillating signal flux Φ_1 (of frequency $\omega_s/2\pi$) through the factor $\cos{[(2\pi\Phi_1)/\Phi_0]}$ provided the factor $J_1 \, [(2\pi\Phi_2)/\Phi_0]$ is not at one of its zeros. [Away from $\Phi_2 = 0$, $J_1(2\pi\Phi_2/\Phi_0)$ approximates to a slightly damped sine function of period Φ_0.]

Now Φ_1 is proportional to the signal current $I_1 = I_s \sin\omega_s t$ coupled to the ring, and with the same factor of proportionality, let I_0 correspond to Φ_0. Then for a given Φ_2, we have from eqn. A8.36

$$V_1' = C_1 + C_2 \cos\left(\frac{2\pi I_1}{I_0}\right) \quad \text{(A8.37)}$$

Now the relative power is proportional to $|V_1'|^2$ and

$$|V_1'|^2 \simeq C_1^2 + 2C_1 C_2 \cos\left(\frac{2\pi I_1}{I_0}\right)$$

if we neglect C_2^2 relative to the other terms. So, as an approximation, the reflected power at frequency $\omega/2\pi$ is of the form

$$P_0 + P_1 \cos\left(\frac{2\pi I_1}{I_0}\right) \qquad (A8.38)$$

where P_1 is small compared with P_0. Because of the non-linear nature of the equations it has not been considered possible to obtain an exact analysis of the dynamic behaviour of the weakly-connected super-conducting ring. Instead, a computer-simulated model[17] and a mechanical analogue[18] have been used to determine the optimum parameters such as the values of L, I_c, coupling factors, and so on.

Finally, the weakly-connected superconducting ring is usually termed a 'squid', (superconducting quantum interference device) and this is a brief and convenient name for it. However, it is not strictly correct, because the term 'interference' applies to the Mercereau effect using *two* Josephson junctions in parallel in a superconducting ring fed by a current. There is no such direct interference of Schrödinger waves in the ring with *one* Josephson junction. There is only flux or fluxoid quantization in such a case. Nevertheless, the term 'squid' is convenient and is likely to stay.

References for appendices

1 MASON, S.J.: 'Feedback theory – some properties of signal flow graphs', *Proc. IRE*, 1953, **41**, pp. 1144–1156

2 LORENS, C.S.: 'A proof of the non-intersecting loop rule for the solution of linear equations by flow graphs' (Research Laboratory of Electronics, MIT, Cambridge, Mass. USA) *Q. Prog. Rept.*, 1956, pp. 97–102

3 MASON, S.J.: 'Feedback theory – further properties of signal flow graphs', *Proc. IRE*, 1956, **44**, pp. 920–926

4 HUNTON, J.K.: 'Analysis of microwave measurement techniques by means of signal flow graphs', *IRE Trans.*, 1960, **MTT-8**, pp. 206–212

5 KUHN, N.: 'Simplified signal flow graph analysis', *Microwave J.*, 1963, **6**, pp. 59–66

6 KERNS, D.M. and BEATTY, R.W.: 'Basic theory of waveguide junctions and introductory microwave network analysis' (Pergamon, 1967), pp. 118–127

7 ADAM, S.F.: 'Microwave theory and applications' (Prentice Hall, 1969), pp. 86–106

8 SORGER, G.U. and WEINSCHEL, B.O.: 'Comparison of deviations from square law for RF crystal diodes and barretters', *IRE Trans.*, 1959, **I-8**, pp. 103–111

9 MEREDITH, R., WARNER, F.L., DAVIS, Q.V., and CLARKE, J.L.: 'Super-heterodyne radiometers for short millimetre wavelengths', *Proc. IEE*, 1964, **111**, pp. 241–256

10 HEISENBERG, W.: 'The physical principles of the quantum theory' (Dover, 1949)

11 FEYNMAN, R.P., LEIGHTON, R.B. and SANDS, M.: 'The Feynman lectures on physics – Vol. 3' (Addison Wesley, 1965)

12 LEECH, J.W.: 'Classical mechanics' (Methuen, 1958)

13 ROSE-INNES, A.C., and RHODERICK, E.H.: 'Introduction to superconductivity' (Pergamon, 1969)

14 PETLEY, B.W.: 'An introduction to the Josephson effects' (Mills and Boon, 1971)

15 SOLYMAR, L.: 'Superconductive tunnelling and applications' (Chapman and Hall, 1972)

16 SILVER, A.H. and ZIMMERMAN, J.E.: 'Quantum states and transitions in weakly connected superconducting rings', *Phys. Rev.*, 1967, **157**, pp. 317–341

17 SIMMONS, M.B. and PARKER, W.H.: 'Analog computer simulation of weakly connected superconducting rings', *J. Appl. Phys.* 1971, **42**, pp. 38–45

18 SULLIVAN, D.B. and ZIMMERMAN, J.E.: 'Mechanical analogs of time-dependent Josephson phenomena', *Am. J. Phys.*, 1971, **39**, pp. 1504–1517

19 GRASSIE, A.D.C.: 'The superconducting state' (Sussex University Press, 1975)

Index

(In addition to subjects, this index contains the names of authors who are mentioned in the text. The names of numerous other authors will be found in the References.)